Lecture Notes in Medical Informatics

Lecture Notes in Medical Informatics

Edited by P. L. Reichertz and D. A. B. Lindberg

33

J. Fox   M. Fieschi   R. Engelbrecht   (Eds.)

AIME 87

European Conference on Artificial Intelligence in Medicine
Marseilles, August 31st – September 3rd 1987
Proceedings

Springer-Verlag

Berlin Heidelberg New York London Paris Tokyo

Editors

John Fox
Biomedical Computing Unit, Imperial Cancer Research Fund Laboratories
Lincoln's Inn Field, London WC2A 3PX, United Kingdom

Marius Fieschi
Laboratoire d'Informatique Médical de la Faculté de Médecine
Université de Marseille
Boulevard Jean Moulin 27, 13385 Marseille Cédex 5, France

Rolf Engelbrecht
MEDIS-Institut, Gesellschaft für Strahlen- und Umweltforschung mbH München
Ingolstädter Landstraße 1, 8042 Neuherberg, Federal Republic of Germany

ISBN-13: 978-3-540-18402-7   e-ISBN-13: 978-3-642-95549-5
DOI: 10.1007/978-3-642-95549-5

2127/3140-543210

Proceedings

AIME 87

European Conference on Artificial Intelligence in Medicine

Marseilles, August 31st - September 3rd 1987

Organized by:

AIME
European Society for Artificial Intelligence in Medicine

Organized in cooperation with:

IIRIAM
International Institute of Robotics and Artificial Intelligence
Marseilles, France

ICRF
Imperial Cancer Research Fund Laboratories, London, UK

GSF-MEDIS
Gesellschaft für Strahlen und Umweltforschung mbH
Institute for Medical Informatics and Health Services
Research, Munich, F. R. Germany

Laboratoire d'Informatique Médical de la Faculté de Médecine
Marseilles, France

Proceedings editors

John Fox,   Marius Fieschi,   Rolf Engelbrecht

Programme Committee

P Adlassnig, Vienna   R Engelbrecht, Munich
M Fieschi, Marseilles   F Gremy, Montpellier
T Groth, Uppsala   A Hasman, Maastricht
A L Rector, Manchester   P L Reichertz, Hannover
P Smets, Brussels   M Stefanelli, Pavia

Organizing Committee

M Fieschi   V Bernadac
P Dujol   B Guisiano
M Joubert   D Riouall
M Roux   G Soula

Tutorial chair

Rolf Engelbrecht, Munich

Local arrangements chair

Marius Fieschi, Marseilles

Programme chair

John Fox, London

REFEREES OF PAPERS

K-P Adlassnig	University of Vienna
P L Alvey	Royal Free Hospital School of Medicine, London
E Carson	The City University, London
S Cerri	Mario Negri Institute, Milan
J R Clough	Thorn EMI Central Research Labs, London
D G Cramp	Royal Free Hospital School of Medicine, London
R Engelbrecht	GSF-MEDIS, Munich
J Fox	Imperial Cancer Research Fund Labs, London
J Fenn	Logica, Cambridge
M Fieschi	Faculte de Medecine, Marseilles
A Glowinski	Imperial Cancer Research Fund Labs, London
F Gremy	C H R Lapeyronie, Montpellier
T Groth	Uppsala University
P Hammond	Imperial College, London
F Harvey	St Thomas's Hospital, London
A Hasman	University of Limburg, Maastricht
J R W Hunter	University of Sussex, Brighton
M Joubert	Faculte de Medecine, Marseilles
M S Leaning	Royal Free Hospital School of Medicine, London
M O'Neil	Imperial Cancer Research Fund Labs, London
E Nicolosi	Royal Free Hospital School of Medicine, London
A Rector	University of Manchester
J L Renaud-Salis	Fondation Bergonie, Bordeaux
P L Reichertz	Institute for Medical Informatics, Hannover
D Sleeman	University of Aberdeen, Aberdeen
P Smets	Universite Libre de Bruxelles
D J Spiegelhalter	MRC Biostatistics Unit, Cambridge
M Stefanelli	Dipartimento di Informatica i Sistemisticag Pavia
H R H Townsend	National Hospital for Nervous Diseases, London
N Walker	Imperial Cancer Research Fund Labs, London
R P Worden	Logica, Cambridge

CONTENTS

KNOWLEDGE ACQUISITION AND REPRESENTATION

MANAGEMENT OF UNCERTAINTY

KNOWLEDGE ENGINEERING TOOLS

GENERAL SESSION

CLINICAL APPLICATIONS (2)

TUTORIAL PROGRAMME
(Not reproduced in proceedings)

Tutorial 1: Acquisition of Knowledge from Medical Databases

Gio C M Wiederhold, M Walker, R L Blum
Stanford University, USA

Tutorial 2: Methods and techniques used in expert systems

Jan L Talmon, Henny P A Boshuizen
University of Limburg, The Netherlands

Tutorial 3: Knowledge Representation

Steen Andreassen, Mike Wellman
University of Aalborg, Denmark
Massachusetts Institute of Technology, USA

**Workshop: The evolution of expert systems
- from MYCIN to ONCOCIN**

Larry Fagan
Stanford University, USA

Methodology

"INTERMED": A MEDICAL LANGUAGE INTERFACE

Christian Mery(*)
Bernard Normier(**)
Antoine Ogonowski(**)

abstract

This paper advocates the use of natural language for querying and updating medical databases. A practical solution is presented, it is centered around "MFRL"-a knowledge representation formalism adequate for the description of medical facts. Underlying MFRL is a certain conceptual view of medical knowledge, which the authors believe to be relevant to a number of applications in Medical AI. A small prototype system embodying all the principles is presented.

I. Introduction

Medical databases used in clinical research or health care management are usually built around nomenclatures (ex. ICD-IX, SNOMED...).

This leads to well known problems such as:

- cumbersome and time consuming access,

- the need of training for correct usage,

- the terms used are of limited specificity, unless very large nomenclatures are used,

- the reasoning capacity is very weak due to reduced semantic organisation.

The solution to the above problems would be to accept the language used by medical personel in their everyday practice.

Therefore, both for querying and indexing, the system should consider as semantically equivalent phrases such as:
"enlarged liver", "hepatic enlargement", "hypertrophy of the liver", "increased liver size", "hepatomegaly", "the size of the liver is increased"...

Furthermore the system should be able to handle qualifying details such as:
"painfully enlarged liver", "hypertrophy of the right lobe of the liver", "severe hepatomegaly", "marked liver enlargement", "the liver is hard and enlarged", "metastatic hepatomegaly"...

Last but not least the system should possess a deductive capacity. For example, one should be able to retrieve "osteosarcoma of the neck of the femur" by queries like: "osteosarcoma of a limb", "bone cancer", "malignant tumor of the lower limb", "primary bone tumor"...

(*)   ROUSSEL UCLAF

        35, Boulevard des Invalides

        75007 PARIS (FRANCE)

(**) ERLI SA

        72, Quai des Carrières

        94220 CHARENTON (FRANCE)

Such deductions have to rely on the the knowledge that "an osteosarcoma is a bone cancer", "a sarcoma is a primary cancer", "the neck of the femur is a part of the upper extremity of the femur", "the femur is a bone of the lower limb"...

Following a long term study of written medical documents, a subsequent conceptual analysis (by Dr. Mery) we have implemented a prototype system (called INTERMED) exhibiting all the above mentionned features.

INTERMED accepts natural language input for both querying and indexing medical facts, translating it into a deep representation expressed in "MFRL".

All the subsequent deductions are made on this deep representation.

This approach allows us to handle multilingual input, since only the parsing process is concerned with the surface language.

II. The MFRL language

MFRL- Medical Facts Representation Language consists of descriptors and connectors organized in a semantic network.

Descriptors represent objects of the "medical world" (organs, biological fluids, lesions, tests ...).

Connectors establish semantic relationships between descriptors (causality, circumstances ....).

II.1 Descriptors in MFRL.

All the descriptors are made up of a semantic class with an attached value.

We have defined a number of semantic classes and subclasses. The following list contains the most important ones (out of 150) with some examples of terminal values:

1. normal constituents of the body
    - organ   TO: liver
    - tissue   TS: bone
    - region   TR: lower limb
    - normal corpuscular elements   EF: leukocyte
    - chemical constituents   CC: calcium

2. nomal functions of the body
    - function   FE: secretion
    - specialized function   FES: respiration

3. characteristics
    - quantitative physical   CP% volume
    - qualitative physical   CQ% color = red
    - nonspecific non-physical   CQN =malignant
    - specific non-physical   CQS =small cell

4. abnormalities
    - lesion   LE = cancer
    - abnormal sound   LFB = murmur
    - of a quantitative physical char.   LECP% volume =increase
    - subjective symptom   LSE= pain

5. changes
    -increase   M = +

6. tests
    - non specific tests   EX = X-ray
    - specific tests   EXS = intravenous pyelography

7. treatments
    - surgery   CHIR= excision
    - medical   MED = aspirin

For reasons explained below, some categories belong to one or more levels of MFRL (for example 5 will not appear above level-1)

The terms used in MFRL may be semantically unidimensional (liver, calcium, arthrosis) or they may be multidimensional (hepatomegaly, hypercalcemia, coxarthrosis...)

A multidimensional term can always be reduced to a combination of unidimensional terms:
hepatomegaly = increase in the size of the liver
coxarthrosis = arthrosis of the hip joint
hypercalcemia = increased calcium level in the blood

II.2 connectors

Connectors establish semantic relationships between descriptors. Following is an excerpt from the 20 connectors used in MFRL:

1. "site" relates an abnormality to a normal component of the body

2. "causing" and "due to" - causal relationship

3. "treated by" relates an abnormality to a treatment

4. "circ" establishes a circumstantial relationship

5. "isite" relates two abnormalities

6. "has" and "is" attach qualifying characteristics to an abnormality or to a normal constituent of the body".

7. "jt" relates two normal constiuents of the body.

II.3 Expressions in MFRL

For a particular natural language statement there corresponds a semantic network in MFRL. Its nodes are composed of descriptors, the connectors being its arcs.

For example, the sentence "hard, painful nodules of the liver due to metastases from a small cell cancer of the right lung" is expressed by the following network:

```
! LE= nodule                                                          !
!   --has-->  CQ % consistency = hard                                 !
!   --causing--> LSE = pain                                           !
!   --site--> TO = liver                                              !
!   --due to--> LE = metastasis                                       !
!               --due to--> LE=cancer                                 !
!                           --site--> TO= lung                        !
!                                     --jt--> DLAT= right              !
!                           --is--> CQS= small cell                   !
!                                                                     !
!_____________________________________________________________________!
```

MFRL is independent of any surface natural language, the words appearing as values in the example above could be replaced by any other unique conceptual labels.

Generally speaking, in the medical world one describes abnormalities (using essentially semantic classes beginning with the letter "L")

Normal facts are thus represented by negating abnormalities. For example "serum calcium is normal" is represented as "there is no abnormality of the calcium blood level".

II.4 Levels of MFRL

The transformation of natural language statements into MFRL formulae is performed through several steps:

1. a "level 0" formula ("MFRL^0") is obtained by fetching instantiated semantic categories or sometimes small networks corresponding to surface words - this is called the "LEX" transformation. for example, "painful (adjective)" will produce "LSE= pain", and "white blood cell" will yield "EF= leukocyte".

2. an "MFRL^0" formula is then subject to transformation by a "general semantic grammar" (i.e. not specific for any particular application, but restricted to the description of medical natural language(***)). The outcome is a "MFRL^1" formula, which is one network containing all the "MFRL^0" information (possibly, slightly modified, for example augmented with causality links).

3. The next step is to transform "MFRL^1" into "MFRL$^{1.5}$", this consists in checking in a Medical Knowledge Base for any additional transformations. For example the network corresponding to "serum calcium at 80 mg/l" will be augmented with the information that it is abnormally low.

III.The INTERMED prototype

INTERMED is a small scale implementation of the abvove principles.

The application domain chosen is- the adverse effects of drugs.
Patients records are input in medical natural language. The system applies all the deductions it can- i.e. saturates the data. (this is close to the approach taken in BDGEN <NICO82>).

(***) In our study of medical texts we have observed some syntactic and semantic variations not found in everyday language.

The MFRL is almost totally transparent to the user, as he communicates with the system in medical natural language.

The system is composed of a "multilingual parser" (supporting English and French-through two different semantic grammars) and an inference mechanism.

It is planned to integrate a simple natural language generator. This task has been analyzed and we can garantee quite acceptable output.
However, currently, during indexation procedure, the present system stores for the original natural language expression with its MFRL formula.

As it has been presented above the system is composed of:

1. a lexicon,

2. two "general semantic" grammars,

3. a medical knowledge base.

Furthermore it includes a number of inference procedures.

Two types of inferences are used:

1. equivalences- establishing a commutative semantic relationship, for example for "metastasis" and "secondary cancer", or "renal excretion" and "urinary excretion".

2. deductions- establishing a non-commutative relationship, like for example the classic specific/generic case: "sarcoma" and "cancer"

III.1 The three functionalities of the system

The three functions that the system performs are:
1. Indexing,

2. Information retrieval,

3. Building the Lexicon.

The first two share in common the three phases of MFRL (i.e. lexical analysis, syntax analysis, and transformations based on the Medical Knowledge Base).

Indexing, continues with subsequent application of all possible deductions and inferences associated with notions present in the indexed phrase (for instance "facial jaundice" will trigger the inferences "increase in the bilirubine blood level").
This has the advantage of considerably speeding up information retrieval, which is the critical task in this type of application.

The network corresponding to an indexed statement is refered to as an "MFRL^2" formula.

Building the Lexicon consists in storing a new word (or expression), its gramatical category and the result of the analysis of its natural language definition. In practice it is not always possible to define new notions in terms of words that are already known, therefore we have included the possibility to resort directly to MFRL expressions.

The user also has the possibility of adding inferences which will fire automatically when this new term appears in a phrase to be indexed.

IV. Conclusions

We have presented a practical approach to the problem of indexing and querying medical textual databases.

Our approach is based on the the use of natural language with all its impacts on the facility of use.

Our system is potentially multilingual (we have examined languages other than English and French), able to deal with a very rich variety of expressions, because all data is stored in a deep internal representation expressed in MFRL.

A large variety of other applications other are possible with INTERMED:

For example an INTERMED database can be easily connected to one or several nomenclature. This can be simply acheived by giving MFRL descriptions of the entries and then performing a series of matches yielding the most appropriate entry.

Such databases cover a considerably large scope of different types of applications, in particular:

1. diagnosis,

2. adverse effects of drugs,

3. summaries of medical records,

Lack of space prohibits an in depth discussion of such proposals in this paper, however we believe that INTERMED can easily be adapted, as an interface to most medical expert systems and medical image bases.

From the technical point of view our approach is somewhat similar to that of Sager <SAGE81> except that we are less concerned with syntactical aspects.

A number of similarities can be seen between INTERMED and systems like BAOBAB (Bonnet <BONN80>) or MEDIUM <JOUB81>. However, all of these lack the "universality" of the semantic classification underlying our MFRL.

V. REFERENCES

<BONN80> Bonnet A.,"Analyse de textes au moyen d'une grammaire sémantique et de schémas. Application à la compréhension de résumés médicaux en langage naturel" Université Pierre et Marie Curie- Paris 6.

<JOUB81> Joubert, "Etude est Réalisation d'un programme de dialogue homme-machine d'aide à la décision en medecine", Thèse d'état, Université d'Aix Marseille II.

<NICO82> Nicolas J.M. and K. Yazdanian "An Outline of BDGEN: A Deductive DBMS", Rapport interne ONERA CERT, Toulouse, Octobre 1982.

<SAGE81> Sager N. "Natural Language Information Processing- A Computer Grammar of English and Its Applications", Addison-Wesley, Reading MA 1981.

INFERENCE ENGINEERING THROUGH PROTOTYPING IN PROLOG

Josef VAN THILLO* and Adolf MULDERS**
* Department of Medical Informatics, Catholic University, Leuven
** St. Elisabeth Hospital, Antwerp, Belgium.

ABSTRACT

Knowledge acquisition is a tedious process when developing medical expert systems. The lack of formal reasoning inherent to medical decision making is partially responsible for this. Prototyping in Prolog, allowing numerous quick successive tests and progressive refinements, is an attractive method for medical
experts as wel as for knowledge engineers to develop an inference engine (IE) for
a particular problem under study. The clinician confronted with a decision
simulating system, rapidly recognizes the most important necessary feature not yet
present. He is then anxious to find out the effect of incorporating the new feature and to apply the revised inference engine on the existing knowledge base
(KB). Independence between the KB and the IE is not only required, but also enhanced by this approach. We applied this technique to build CASETIP (a Computer
Advice System for the Empiric Therapy in Pneumonia) using Micro-Prolog on a 256
Kbyte microprocessor.

INTRODUCTION

Acute pneumonia remains an important and challenging problem in modern medicine
(1). Proper antibiotic therapy must be instituted prior to, and later also often
without any definite identification of the causal organism. The recognition of
pneumonia syndromes helps to narrow the spectrum of possible etiologic agents and,
as a result, helps the clinician to develop a rational regimen of antibiotic
therapy. Because most non-specialized clinicians are unfamiliar with the required
knowledge to achieve this goal, they are more likely to use some aspecific treatment schedules based upon the severity rather than the specificity of the patient's condition. However, a specific therapy enhances the efficacy and reduces
the risk of side effects as well as the cost of treatment.
The problem of pneumonia therapy is thus a valid candidate for support by computer
technology. We will demonstrate that even limited hardware and software resources
can support this specific task.

No general formalism exists for medical decision making (2). Most decision support systems develop their own ad hoc solutions. This may be regrettable at first sight. However, it follows logically from the diversity of medical decision making paradigms (3). The difficulties with knowledge acquisition (KA) often result from a lack of authenticity of the imposed decision strategy. A highly flexible **"inference engineering"** method can ease the KA process.
We applied **rapid prototyping in the presence of the expert** (4) as such a flexible decision engineering technique, to build CASETIP (a Computer Advice System for the Empiric Therapy In Pneumonia). We found that Prolog is a suitable tool for this task (5,6).

METHODS AND RESULTS

The main participants of the project are a practicing family doctor with special training in informatics (JVT) and a practicing pneumologist with special interest in acute infectious diseases (AVM).

The three main stages of the project were :
1. a search of the medical literature about pneumonia and related problems,
2. the development and testing of knowledge base and inference engine,
3. the validation.

This paper describes the second stage. The validation is still going on. About 30 retrospective pneumonia cases were used to debug the system s behaviour. At present a number of cases that is statistically sufficient for a real validation is being collected in a prospective way in three different environments : in a general practice environment, on a pneumological ward of a regional hospital as well as in a university hospital.

As hardware we used a Tulip System PC (IBM compatible) with 256 KB RAM and two diskette drives, and as software Micro-Prolog 3.05 running under MS-DOS 3.10.

We now describe the major steps in the development of the inference engine of CASETIP. As a first step, we defined the basic diagnostic categories (further called "pneumonia types" or "types" for short), as well as a series of diagnostic criteria (further called "signs), and some treatment schedules. The following naive reasoning scheme was programmed in Prolog :

$$\text{sign} \ll -- \gg \text{type} \ll -- \gg \text{treatment}$$
$$\text{DIAGNOSIS} \qquad \text{THERAPY}$$

The most striking deficiency here was the dichotomic (yes/no) definition of predictive values. This led to the identification (from the scientific literature and from personal experience) of **obligatory, suggestive, non-specific** and **exclusionary** signs, arranged in criteria table-like patterns.

Atypical pneumonia e.g. is characterized as follows.

- Suggestive signs are : age under 30, irritative cough, acute onset, headache and/or myalgia, high fever, community-acquired.
- Obligatory signs are : high fever and community-acquired.
- A non-specific sign is : community-acquired.
- Exclusionary signs are : cavitations on chest X-ray, putrid sputum and high fever lasting for more than one week.

A diagnosis is included in the working differential diagnosis (DD) if :

1. There is at least one suggestive sign,
2. All obligatory signs are present,
3. More than only non-specific signs are present,
4. There is no exclusionary sign.

The resulting system behaviour clearly demonstrated that the criteria should be used and interpreted in different ways, according to their **sensitivity and specificity** at one hand, and to their **availability** at the other. This led to a diagnostic process in **two stages**. There is, firstly, a broadly orienting stage, inductively evoking all the diagnoses to be considered, based upon sensitive and easily available criteria. This stage leads to a basic differential diagnosis (DD). In the second, deductive (hypothesis driven) stage, the DD is progressively refined using more specific criteria and, if needed, also the results from further investigations.

Some cases were overdiagnosed : part of the performed diagnostic work up turned out to be without any therapeutic implication because one single treatment could have covered the whole DD at an earlier stage. Thus, the program was redesigned so as to make use of this simplification. Further differentiation e.g. between mycoplasma pneumonia and pneumococal pneumonia isn t needed because both respond to erythromycin, for example.

In all other cases the system tried to reach a unique diagnosis. Sometimes, however, it is necessary to use poorly reliable diagnostic criteria in the final diagnostic steps. A better alternative was to stop the diagnostic process at an intermediate stage, and to propose a combination therapy. Combining therapies doesn't mean a simple adding up of two or more antibiotics : combinations must be evaluated and, if needed, be revised according to two criteria : compatibility and non-redundancy. The combination of ampicillin and cefotaxime e.g. can be reduced to cefotaxime alone, because the spectrum of the former is included in that of the latter.

The two latter features are examples of the non-dissociation of diagnosis and therapy in the second (hypothesis driven) stage.

Now our attention was drawn to an even more important illustration of this non-dissociation principle. When instituting a therapy, the clinician must point out within what delay he expects significant improvement. When improvement is not observed within this period, one must conclude that therapy has failed. For some therapy failure, the simple and logical solution is the institution of an alternative medication. However, **therapy failure** may also be an unexpected outcome. This should be considered **as new diagnostic information**, raising doubt about the original diagnosis. This was solved by characterizing each type in the knowledge base as, either, possibly therapy resistant, or, invariably responsive. Alternative therapies are provided only for the former category. Therapy failure permits the invariably responsive types to be thrown out from the current DD.

In rare instances, it happens that a so-called invariably responsive pneumonia type may not respond to therapy, so we are left with an empty DD. This is not unusual in clinical practice : it must be interpreted as a reliable argument against the diagnostic hypothesis and therefore against the way the diagnosis was derived. In such cases, the program repeats the second (deductive) stage, disregarding the rules applied so far. The therapy which failed will be registered in the patient record. The first (inductive) stage is never repeated automatically, although the user is free to check and reconsider the input data, and eventually start over the whole session.

At first sight, it seems unreasonable to disregard **diagnostic knowledge** that has been used just at a previous stage. But again, this is quite common in clinical practice. Informally, one could state that every physician is aware of the relative nature of his knowledge. Not any medical rule is universally applicable, and the doctor should be able to disregard part of his reasoning scheme in some particular cases. Formally stated in Bayesian terms : the positive predictive value of any criterion depends upon the a priori probability of the disease. The estimation of this prior probability (and thus of the reliability of some diagnostic criterion) varies during the diagnostic process, and may be dramatically reduced at the end of it, e.g. when therapy fails for a so-called invariably responsive disease. Specific criteria that seemed reliable at an earlier diagnostic stage (when it was not known that the prior probability was very low), become useless and must be disregarded a posteriori.

At times, this procedure will not yield an acceptable solution. Then the program will fail and consider itself as unqualified to solve the problem under study. Indeed, CASETIP isn't meant to solve the most complex pneumonia cases nor to replace the specialised clinician. It's goal is to progagate some degree of expert knowledge to a larger group of ad hoc non-specialised physicians, who are familiar with evaluating external advice and integrating it in a broader decision making context.

DISCUSSION

Empiric therapy for pneumonia shows several principles of medical decision making. The experienced clinician, however, isn't always conscious of the decision principles he uses in routine practice, and finds it hard to make them explicit. We found that confronting the expert with a prototype decision making system is a pretty direct way to reveal the missing features.

CASETIP has been tested with about 30 cases, some of which were artificially converted into more complicated ones, in order to force the deduction process into untrivial branches. Although this number isn't large enough to make any statistically reliable conclusion about the validity of the system, it suffices to explore almost all the possible deduction sequences and cycles. For each case, we went through the session in "full explanation mode", which means that every atomic deduction step is made explicit and commented upon. Although, too time consuming for routine use, this mode is higly suited for the "dissection" we aimed at. This procedure couldn't reveal illogical or unnatural reasoning for any of the cases studied. This permits us to conclude that the described process during which a medical reasoning strategy is progressively refined through iterative prototyping in Prolog, is a usefull method for the design of AIM systems with higher user acceptance. On the other hand, it is a rich source of insight in the fundamentals of medical decision making. We like the term "inference engineering" instead of "knowledge engineering" to describe this technique in order to highlight the emphasis on unraveling the dynamic process of clinical reasoning rather than the formalism of static medical knowledge.

We acknowledge Prof. J.L. Willems and Dr. L.J. Dekeyser (University Hospital Gasthuisberg, Department of Medical Informatics, Leuven, Belgium) for the stimulating discussions.

REFERENCES

(1) G.R. Donowitz, G.L. Mandell, Empiric Therapy for Pneumonia. Reviews of Infectious Diseases, 5 : S40 - S48, 1983.

(2) H.E. Pople, Jr. Heuristic Methods for Imposing Structure on Illstructured Problems : The Structuring of Medical Diagnosics. In : "Artificial Intelligence in Medicine." (Ed. Peter Szolovits). Westview Press, Boulder, Colorada 80301.

(3) A.S. Elstein, L.A. Shulman, S.A. Sprakfa. Medical Problem Solving : An Analysis of Clinical Reasoning. Harvard University Press, Cambridge, Mass. 1978.

(4) J. Fox et. al., Knowledge Acquisition for Expert Systems : Experiences in Leukemia Diagnosis., Methods of Information in Medicine, 24 # 2, April 1985, pp. 65-72.

(5) R.A. Kowalski, Logic for Problem Solving. Elsevier North Holland, New York, 1979.

(6) L. Sterling, E. Shapiro, The Art of Prolog : Advanced Programming Techniques. The MIT Press, Cambridge, London.

The evaluation of clinical decision support systems: a discussion of the methodology used in the ACORN project

Author: Dr Jeremy Wyatt, Dept. of Thoracic Medicine,
          Westminster Hospital, Dean Ryle St, London SW1P 2AP.

Summary

The evaluation of medical expert systems, particularly those intended for decision support in the clinical domain, has not received sufficient emphasis. The techniques used for the evaluation of 14 medical systems are reviewed and contrasted with those used for the evaluation of new drugs. A three stage procedure is outlined. This is currently being used for the evaluation of ACORN, a decision support system to aid the admission of patients to cardiac care units.

Introduction

a) Why evaluate ?

There are only three reasons for using medical knowledge to build an expert system: to explore a knowledge engineering problem [Aikens '83, Fox et alii '80], to build a system to fill a practical medical role [Aikens et al '83, Weiss et alii '81] or to transfer knowledge to an expert system as a form of 'rare skills archiving' [Alvey et al '86]. However, some medical systems fall into none of these categories, and are analogous to 'me too' drugs. It is these that risk the criticism from doctors and funding agencies that could herald the dreaded 'AI backlash'.

The history of science reveals that it is measurement which marks the watershed in most disciplines, replacing speculation as the means of determining whether theories work or not. So it is with the expert systems field: to transform it from a mass of speculative theories and prototypes, many never implemented, we need to measure what our systems are doing. The researcher can then evaluate progress against real targets, a purchaser can choose between rival systems, and all of us can claim that we are working in a scientific area.

In medicine, performance evaluation is especially important if the safety of our patients is to be entrusted to expert systems. Doctors, notorious for their mistrust of computers, will need convincing that such systems can offer them anything other than an assault on their integrity and clinical freedom. Responsibility is another key isssue: not only will the performance of a medical expert system be important in deciding whether a doctor was right to use one in a case of alleged negligence, but, if an American judgement of 1981 is upheld, a doctor may be found negligent if he does <u>not</u> use a computer system that has been shown to improve performance.

b) What to evaluate ?

There are probably as many motives for evaluating a system as there are system builders. Sometimes the investigator is not as interested in the question 'Does it work ?' as 'How does it work ?' - the 'pragmatic' and 'explanatory' motives for evaluation, respectively [Spiegelhalter '83]. The true purpose for performing an evaluation must be clear from the outset, as this dictates what is measured.

In medicine, unlike some application areas, there are many measurements that can be made, only a few of which are useful. For example, the diagnosis is a precursor to determining treatment and investigations; these in turn determine the immediate outcome (treatment success or failure, referral, hospital admission and length of stay), but this is only a snapshot of the disease process, and the real issues are whether the patient survives, if so for how long, and at what cost to themselves (quality of life) and to society (days off work, use of resources). Thus, an expert system that substantially improves the accuracy of diagnosis of patients attending a hospital with chest pain may have a negative effect on survival if it also slows down the patients' admission to a cardiac care unit.

There are, therefore, other factors that must be evaluated. Some, such as the speed with which decisions are made, the reliability of the system in terms of down-time, and the amount and quality of data that the system uses to make its decision [Fox et alii '80], can be measured. Others, such as the satisfaction of the users with the system and the transparency and maintainability of the knowledge base are currently difficult to evaluate.

c) The current position

The evaluation of process for decision support systems was first clearly enumerated in 1975 [Shortliffe et alii '75], and reiterated in more recent publications [Szolovits et al '82, Gaschnig et alii '83, , Spiegelhalter '83, Wasson et alii '85]. Despite this, in a cursory review of 14 papers on artificial intelligence in medicine (table 1), only three adequately described the system's clinical role, seven had a scientific test set analysis, and one was subjected to a field trial. Sadly, only two of these 14 systems, PUFF and EXPERT, are currently in use, both for the interpretation of laboratory results.

These three key areas of performance evaluation, namely the clinical role definition, test set analysis and the field trial, therefore seem the most appropriate to explore further.

Performance evaluation, stage 1: Defining the system's clinical role

It is easy to forget that an adequate description of the system and its role must be given before evaluation can start.

This should consist of a problem specification with a protocol for how the correct or 'gold standard' solution to the problem is to be derived (see next section), of the potential users of the system and their current performance at the task, of the amount, quality and types of data available for input, of the output required, and the level of explanations needed. In addition, the 'clinical context' should be described: this includes the area where the system will be used (GP surgery, A/E dept., etc.), the types of patient and the clinical implications of the system's advice. A further requirement is for a judgement to be made of the significance of errors, to enable a table of 'utilities' or weightings for correct and incorrect decisions to be drawn up.

In many clinical contexts, the objective is not to ensure that every decision is made at expert level [cf. Shortliffe et alii '84], but simply to improve the quality of current decisions. This means that the current level of the doctors' decision making accuracy (and speed if appropriate) needs to be determined. Studying the doctors may itself alter their performance - this is the Hawthorne Effect, analogous to Heisenberg's Uncertainty Principle in the field of quantum physics, which states that any observation carried out on a particle disturbs that particle. To eliminate the Hawthorne Effect, the doctors should be unaware that they are being studied - this may take some ingenuity !

Example: the ACORN project

ACORN (an acronym for Admit Cardiac care unit OR Not) is a system that we are developing to give rapid, accurate advice to doctors managing patients attending an accident and emergency department (A & E dept.) with chest pain. The problem is that the current decisions are often innaccurate and patients wait too long in the A & E dept. The 'correct' management decisions are determined by a concensus of 2 out of 3 independent assessors after follow-up in a chest pain clinic or on the wards. The system is designed for the casualty nurses to input data from a nurses' questionnaire and a doctors' ECG reporting form (accuracies evaluated by repeatability and comparison studies); the system then gives advice about type of underlying problem, a management recommendation (admit or not), and whether an ECG or CXR is required. No explanations are currently given, as the doctors review the input data and no educational role is intended yet; the system's advice about admission and investigations is treated more like the results of a special test. The utilities are that a correct decision to admit a patient with an acute MI to the CCU scores +20, a decision to discharge this patient home would score -20, and a decision to admit a patient to the CCU who could have gone home scores -5.

Performance evaluation stage 2: The Paper Test

It has been said [Wasson '85] that 'clinical prediction rules are a form of health care technology, and should be studied with the same care as new drugs'. This suggests that a comparison between the evaluation techniques for new drugs and medical expert systems could be fruitful (see figures 1 & 2).

In drug evaluation, the role definition corresponds to the definition of a disease and its pathophysiology. The next stage is then to simulate the disease in laboratory animals and to assess the proposed agent's efficacy and toxicity. This simulation is analogous to the refine and test cycle of a decision support tool, using a 'training set'. In drug evaluation, it is the results of the next stage, testing the drug on human volunteers, which determine whether a full clinical trial can go ahead. Similarly, it is the results of a 'test set' analysis that determine whether a decision support tool is suitable for a field trial. In drug trials, the format of these analyses is specified and monitored by an agency such as the Committee on Safety of Medicines (CSM). With decision support systems, the design of both test set analyses and field tests is left to the investigators: as a result, accidental bias and poor control can occur, particularly in certain areas.

A test set is a collection of data which is distinct from the training set, randomly drawn from the population, to which the investigators have had no previous access. It should be large enough to contain several examples of each of the problems that the expert system is expected to solve. For a statistically-based decision aid [Wasson et alii '85] the test set should contain a minimum of 5 cases in each class per item of data input. Thus, if one is diagnosing tuberculosis from 20 items of data, there should be at least 100 patients with proven TB in the test set.

Obtaining the correct or 'gold standard' answer to a medical problem is often difficult. Strictly, if it is a diagnosis that is at issue, only a post mortem is satisfactory, as a recent study comparing clinical diagnosis on inpatients with post mortem results revealed an overall clinical accuracy of only 47%, with serious treatable diseases missed in 13% [Mercer et al '85]. However, histology is often not available and a follow-up study with appropriate investigations must be used. An alternative, used in some studies [Yu '79, Alvey '86], is to give one or more experts copies of the data input to the expert system and take their view as correct, without attempting to find out what was the actual solution to the problem. In this case, an assessment of the experts' repeatability is valuable [Fox et alii '85].

In the ACORN study, we enforce a minimum two month follow-up period, use diagnostic criteria from a published source, and the multiple independent assessors are blinded to the source of any diagnosis they are called on to verify, and to whether the patient is in control or computer groups. If there is remaining doubt, the patient is either recalled for further tests or excluded from the study.

In studies that report the results of a test set, it is often the number of agreements with the expert that are specified. It is more informative to include a 2 X 2 table (see figure 3) as from it, the false negative and false positive rates [Wulff '76] can be calculated, and these, combined with the utility structure described above, make the contribution of the expert system clearer.

<u>FIGURE 3:   2 X 2 table, with measures derived from it.</u>

| | | 'Gold Standard' | | Totals |
		True	False	
E S x y p s e t r e t m	True	a	b	a+b
	False	c	d	c+d
Totals:		a+c	b+d	n

Crude 'Accuracy'   = a+d/n

False negative rate   = c/a+c

False positive rate   = b/b+d

Predictive value pos. = a/a+b

Predictive value neg. = d/c+d

The main aim of the 'paper evaluation' outlined above is to establish whether the expert system meets the predefined targets of accuracy and utility to justify a 'field test' from the ethical point of view, or whether the refine and test cycle needs to be repeated. In addition, it will establish the feasibility of conducting a field trial, by clarifying the time the system takes to reach a decision, the amount of data and number of cases required, and if the users will be able and willing to use the system on at least half of the cases they see.

Performance evaluation stage 3: The Field Test

The fundamental question to answer about any new medical technique or drug is 'Does it make a significant difference to the patients, the staff looking after them or the resources which are used in the process ?' This requires a field test or clinical trial. For drug evaluation, this is the familiar prospective, double blind, randomised, placebo controlled, clinical trial [Armitage '71]. For a medical decision aid, one can view the trial as a 'Turing Test' [Spiegelhalter '83], with a blinded expert comparing the performance of the computer (or the doctor aided by the computer) with that of the doctor unaided. Surprisingly few decision support systems have ever been submitted to a field test, R1 and the Leeds abdominal pain system being exceptions.

Before the trial can be planned, the correct size of the

'sampling unit' must be determined [Spiegelhalter '83]. If a system is designed to have an educational effect on its users, to provide feedback or 'critiquing' of plans [Miller '83], then doctors or even firms should be allocated to the control and intervention groups, while if it has a direct problem solving role, then allocation of patients is more appropriate.

Permission must be sought of an impartial ethics committee before starting any trial involving patients, In addition, permission should probably be sought of the patients, even when it is the performance of the doctor or firm which is being studied, as their treatment may be altered as a result [Wyatt et alii '86]. Another view is that, since doctors treat patients differently and patients take their luck as to which doctor they see, if a doctor has an expert system available to him then the patient need not give consent as he will probably receive a higher standard of care. The question of consent in clinical trials of decision support systems remains a grey area.

In a controlled trial it is vital to ensure that the control and computer groups differ only in the use of the computer to aid decision making in the latter group; this means that the controls should be simultaneous and randomised. It must be possible to conceal from the investigators which group an individual patient is in. In the ACORN trial, patients are randomised by casualty nurses using a sealed envelope, and a 'control' or 'computer' label is put on one sheet of the hospital notes. An assistant then ensures that the investigators only see notes that do not contain this labelled sheet.

The use of an expert system will alter patient management, and any 'side effects' may reduce the value of the decision aid. Some will be obvious from the role definition above, while others will have been identified during the running of the test set. Some, however, such as the patients' response to the decision aid, may only be apparent during the field test. Flexibility must be allowed in the definition of possible side effects; the controlled trial structure then allows a comparison to be made between expert system and control groups.

It is always tempting, particularly when deadlines are approaching, to curtail a trial when it appears to show a successful result, but the size of the trial must be determined in advance, by consideration of the desired confidence limits for the result, the expected improvement due to the use of the expert system and the scatter of the parameters being measured. This calculation in particular, and trial design in general, is an example of the contribution that statisticians can make.

After a field trial, it would be easy to conclude that the value of the expert system is now proved, but there are limitations on how widely the conclusions can be applied. For example, an expert system that is effective at speeding up the admission of adults with chest pain to a cardiac care unit (CCU) cannot be used for children, nor in a hospital where there is no CCU. Some would argue [Spiegelhalter '83] that a multicentre trial is needed to control for the NIH (Not Invented Here) Effect - the success of a system in one hospital may be due solely to the enthusiasm of the investigating team. Advances in medical knowledge may rapidly make a system obsolescent; unless a knowledge base has been carefully planned for ease of updating [Fox et alii '86], an apparently minor change may require a radical revision, and a new evaluation cycle is then necessary.

Conclusions

The   scientific evaluation of a decision support tool requires three distinct stages, all of which are time consuming, but which allow a definite statement to be made about the contribution of the system to medical practice.   Careful evaluation of medical expert systems is a vital part of the development of working systems that will offer enormous benefits to patients and physicians alike, and will help to place expert systems in particular, and AI in general, on a sounder footing.

References

1. Adams ID, Chan M, Clifford PC et alii, 1986: 'Computer aided
   diagnosis of acute abdominal pain: a multicentre study'; BMJ
   vol. 293, pp. 800-804.
2. Aikins JS, Kunz JC, Shortliffe EH, 1983: 'PUFF: an expert
   system for the interpretation of pulmonary function data';
   Computers & Biomedical Res. vol. 16, pp. 199-208.
3. Aikins JS, 1983: 'Prototypical knowledge for expert systems';
   Artificial Intelligence vol. 20, pp.163-210.
4. Alvey P, Greaves MF, 1986: 'Observations on the development of
   a high performance system for leukaemia diagnosis'; in Proc.
   Expert Systems '86 (Brighton), ed. Bramer M., pub. Cambridge
   University Press.
5. Armitage P, 1971: 'Statistical methods in medical research';
   pub. Blackwell Scientific, Oxford.
6. Catanzerite VA, Greenburg AG, Bremerman HJ, 1982:' Computer
   consultation in neurology: subjective and objective
   evaluations of the Neurologist system'; Comput. Biol. Med.
   vol. 12, pp. 342-355.
7. Diamond GA, Staniloff HM, Forester JS et alii, 1983: 'Computer
   assisted diagnosis in the noninvasive evaluation of patients
   with suspected coronary artery disease'; J Am Coll Cardiol.
   vol.1, pp. 444-455.
8. Ellam SV & Maisey MN, 1986: 'A knowledge based system to assist
   in medical image interpretation: design and evaluation
   methodology'; in Proc. Expert Systems '86 (Brighton), ed.
   Bramer M.,pub. CUP.
9. Fieschi M, Joubert M, Fieschi D et alii, 1983:' A program for
   expert diagnosis and therapeutic decision'; Med Inform. vol.
   8, pp. 127-135.
10. Fox J, Barber D, Bardhan KD, 1980: 'Alternatives to Bayes: a
   quantitative comparison with rule-based diagnostic
   inference'; Meth. Inform. Med. vol. 19, pp.210-215.
11. Fox J, Myers CD, Greaves MF, Pegram S, 1985: 'Knowledge

acquisition for expert systems: experience in leukaemia diagnosis'; Meth. Inf. Med. vol. 24, pp. 65-72.

12. Fox J, Duncan TD, Frost D, Glowinski A, Hajnal S, O'Neill M, F, 1986: 'Organising a large knowledge base: the Oxford System of Medicine', paper presented at Expert Systems '86 (Brighton).

13. Gaschnig J, Klahr P, Pople H, Shortliffe E, Terry A, 1983: 'Evaluation of expert systems: issues and case studies'; in Hayes-Roth F, Waterman DA & Lenat D (eds), 'Building expert systems', Addison Wesley 1983.

14. Hudson DL, Cohen ME, Deedwania PC, 1984:' Emerge: a rule based expert system implemented on a microcomputer'; Microcomputer Applications vol. 3, pp. 79-83.

15. Mercer J, Talbot IC, 1985: 'Clinical diagnosis: a post-mortem assessment of accuracy in the 1980s'; Postgrad. Med. J. vol. 61, pp. 713-716.

16. Miller PL, 1983: 'Medical plan-analysis: the ATTENDING system'; proc. IJCAI-83, pp. 239-241.

17. Politakis P, Weiss SM, 1982:'A system for empirical experimentation with expert knowledge'; Proc. 15th Hawaii Int. Conf. on Systems Science, 2, pp. 649-657.

18. Schwartz WB, Patil RS, Szolovits P, 1987: 'Artificial intelligence in medicine: where do we stand ?', NEJMed vol. 316, pp. 685-688.

19. Shortliffe EH, Davis R, Axline SG et alii, 1975: 'Computer based consultations in clinical therapeutics... the MYCIN system'; Computers & Biomed. Res. vol. 8, pp. 303-320

20. Shortliffe EH, Clancey WJ, 1984: 'Anticipating the second decade'; in 'Readings in medical AI', eds. Shortliffe & Clancey, pub. Addison Wesley 1984.

21. Spiegelhalter DJ 1983: 'Evaluation of clinical decision aids, with an application to a system for dyspepsia'; Statistics in Med vol.2, pp.207-216.

22. Szolovits P, Long WJ, 1982: 'The development of clinical expertise in the computer'; chap. 4 in 'Artificial inteligence in medicine', ed. Szolovits P, pub. Westview, for AAAS, Washington.

23. Wasson JH, Sox HC, Neff RK & Goldman L: 1985: 'Clinical prediction rules: applications and methodological standards'; NEJMed vol.313, pp.793-799.

24. Weiss SM, Kulikowski CA, Galen RS, 1981: 'Developing microprocessor-based expert models for instrument interpretation'; in Proc. of seventh IJCAI, pp. 853-855, pub. Kaufmann Inc, Los Altos.

25. Weiss SM, Kulikowski CA, Amarel S, 1978:' A model-based method for computer-aided medical decision-making'; Artificial Intelligence vol.11, pp. 145-172.

26. Wulff HR, 1976: 'Rational diagnosis and treatment'; pub. Blackwell Scientific Publications, Oxford.

27. Wyatt JC, Emerson PA, Crichton NJ, 1986: 'Computer aided diagnosis of acute abdominal pain (letter) ; BMJ vol. 293, p. 1305.

28. Yu VL, Fagan LM, Wraith SM et alii, 1979: 'Antimicrobial selection by computer: a blinded evaluation by infectious disease experts'; JAMA. vol. 242, pp. 1279-1282.

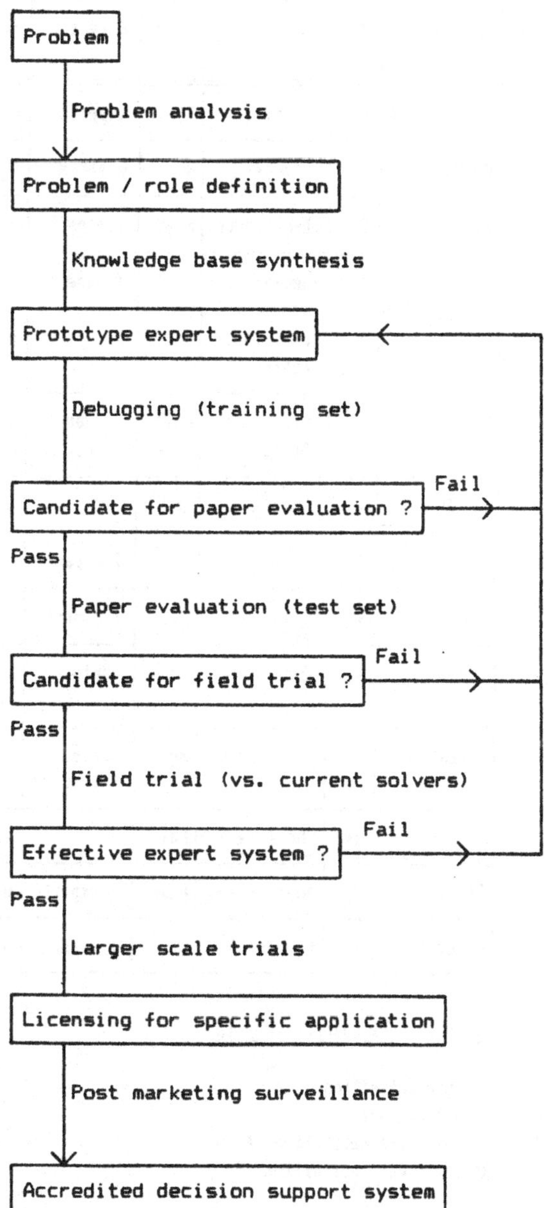

FIGURE 1: Drug development cycle

Disease

Analysis of pathophysiology

Disease process definition

Drug synthesis

Prototype drug

Pharmacokinetics, toxicity tests
(animals)

Candidate for clinical testing ?   Fail

Pass

Clinical testing (volunteers)

Candidate for clinical trial ?   Fail

Pass

Small scale trial (patients)

Effective drug ?   Fail

Pass

Larger scale trials

Licensing for specified conditions

Post marketing surveillance,
further trials.

Accredited drug

FIGURE 2: Proposed cycle of medical expert
system development

Problem

Problem analysis

Problem / role definition

Knowledge base synthesis

Prototype expert system

Debugging (training set)

Candidate for paper evaluation ?   Fail

Pass

Paper evaluation (test set)

Candidate for field trial ?   Fail

Pass

Field trial (vs. current solvers)

Effective expert system ?   Fail

Pass

Larger scale trials

Licensing for specific application

Post marketing surveillance

Accredited decision support system

TABLE 1: Analysis of 13 reported medical expert systems, using classification from text.

System name	Ref. no.	Domain	Role def.	Gold st.	Test set	Field test	Comment
NEUROLOGIST	6	Neurology	Good	Experts	n=30	'observe'	Stated goals, not blinded
CASNET	25	Glaucoma	No	Experts	No	Observed	K/E methodology
EXPERT	24	Electrophor.	Fair	1 expert	n=256	In use	Lab. instrument
SEEK	17	Rheumatol.	Fair	3 experts	n=121	No	Paper about knowledge acquisition
PUFF	2	Pulm. func. tests	Good	2 experts	n=144	In use	Agreements quoted.
CENTAUR	3	Pulm. func. tests	Fair	2 experts & PUFF !	n=100	No	Paper about prototypes for knowledge representation
PSYCO	10	Dyspepsia	No	?	n=50	No	Paper compares Bayes & IKBS
EMYCIN	11	Leukaemia	No	1 expert	n=100	No	Checked experts repeatability
SPHINX	9	Diabetes	Fair	1 expert	n=50	No	Paper about H & T
Leeds	1	Abdo. pain	Good	various	other paper	n=12600	Trial good but historical controls & limited follow-up
CADENZA	7	Cor. H.dis.	No	Angio,1yr follow-up	n=1000	No	Bayesian type system
EMERGE	14	Chest pain	Fair	?	No	No	Tables of disk space needed only
Ellam's	8	Thyr. scans	Fair	1 expert	n=100	No	Expert entered scan data
Alvey's	4	Leukaemia	No	1 expert	n=400	No	100% accuracy. ? clinical role.

<u>Summary of evaluation results</u>

<u>a) Role definition:</u>
No definition: 5
Partial role definition: 6
Adequate role definition: 3

<u>b) Test set criteria matched:</u>
No test set : 2 (1 reported extensively elsewhere)
Major criticism: 4 (no gold standard 3, inadequate size 1)
Minor criticisms: 7 ('accuracy' only given, poor gold standard)

<u>c) Field test:</u>
No field test: 9
Subjective field test: 2
Objective field test: 1 (historical controls, limited follow-up)
In use: 2

Matching Patients: An Approach for Decision Support in Liver Transplantation

G. Tusch, J. Bernauer, P.L. Reichertz

(Medical School Hannover, F.R.G.)

ABSTRACT

Clinical research often is based on matching of identical patterns in patient courses. The LTX system is designed to support clinical research and decisions for the post-operative care of liver transplant patients by identifying "similar" patient courses. The basic idea is a frame representation of each day in a patient's course including historical information. On top of this we constructed a time oriented language to represent clinical concepts like "high bilirubin level over 4 days".

The system is implemented on a PC, where an integration into the hospital information system is essential (identification via MHH-identification number, connection to the lab computer to get the patients' lab data, etc.).

1. INTRODUCTION

In the postoperative care of liver transplant patients the clinician has to be aware of many problems affecting the transplanted organ. The most important ones are rejection, infection, toxicity, bile duct obstruction, or problems in the arterial supply. Many of these problems can occur concurrently, so that the physicion has to take into account a large amount of parameters to aequately deal with the variability of the patients course.

In 1985 there we started at Hannover Medical School the development of a prospective patient registry, part of which was the documentation of the postoperative phase in the care of liver transplant patients. The basic idea was that decision support in this complex area is only possible when there is a reliable documentation of the patients' courses in the beginning. This patient registry has three goals:

- to serve as a basis for the evaluation of the patient courses (developing a "Gold Standard")

- extraction of groups of patients with defined parameter values for specific clinical questions

- searching for "similar" constellations (patterns) in the patients courses to support research (exploratory data analysis) and later on decisions in specific situations in a patient's clinical course.

One possible approach in clinical research is to identify similar patterns in patients' courses and observe the outcome of these situations. In the first phases of this kind of analysis "similarity" is an intuitive concept derived from clinical experience and known pathophysiological models. Formalizing this intuitive concept of similarity by reducing it to measurable clinical parameter values is the first step in this kind of explorative analysis. This is an iterative process, where you always have to test if the selected set of parameter values is sufficient and exhaustive to describe the specific concept of similarity you have in mind. Mathematically seen a measure of similarity is a distance measure that in simple cases can be realized as a continous function (e.g. power transformation (see -6-), and in complex cases can be a whole complex program that incorporates clinical and pathophysiological concepts (as in our case).

This iterative process is exactly what knowledge engeneering for expert systems means. So we started to develop an expert system to support the identification of "similar" situations in patient courses. The goal was in the first step to support clinical research, and in the second to support a clinical decision situation in a "pre-statistical" manner in cases where no sufficient statistical information (risk scores etc.) is avaible.

2.   BASIC IDEAS OF THE LTX SYSTEM

In the first step we developed a system that does use a simmple distance measure: pattern matching. "Similarity" in this concept is the exact match of a pattern of a clinical situation. We plan to use more sophisticated measures.

To clarify our ideas we will contrast these with similar approaches in expert system research. The classical problem in MYCIN-type systems is the single consultation of a non-expert user to get a definite recommendation. This idea is continued in the ONCOCIN system (-9-), where the time-consuming dialog for information gathering is shortened by direct acces to a clinical patient file that is used for documentation purposes. Here the expert system supports clinical trials by checking and utilizing the procedural knowledge for the correct performance of the study, and recommends dose adjustment according to the cancer trial protocol. The clinical knowledge for cancer treatment has been established by many trials, and cancer study design is a well developed field in statistics, so here randomized clinical trials are appropriate. Surgical trials are much more difficult to handle with classical statistical tools than a randomized clinical trial than therapeutic trials because there are a lot of more sources of biases, as the improving surgical skill and performance of a new developed surgical procedure, or because random selection of patients is unethical (preference of urgent cases), etc. (see -10-,-2-).

So, when there is no indication for a randomized clinical trial, the problems are identification of "typical" patterns and courses of the postoperative complications, determination of the predictive value of parameter patterns, or development of risk scores for typical situations. The risk scores are use to select the appropriate treatment, or intervene at the appropriate time.

The basis of our approach is the following simple model, how this research process in practical situations looks like:

1. selection of a problem area

2. Identification of similar patients courses or similar parts of them

    This similarity in general is an intuitive definition that involves clinical expertise, clinical concepts as "high bilirubin level", and some statistical results from the literature.

3. exploratory data analysis

    Extractraction of important features and showing that these discriminate to other cases.
    Here we need a "Gold Standard" to evaluate the outcome of the different situations and look for correlations.

4. Development of a score (risk etc.)

5. Test the score in a prospective trial (sufficient number of patients against the "Gold Standard")

6. Use the score for patient treatment

The clinical <u>decision</u> process is a short version of this research process. Considering a patient's course at a critical point means going up to step 3, to decide what clinical action or treatment is appropriate compared with the outcomes of "similar cases".

The part of the clinical research/decision process that we want to support is the second one in the list above. This situation has some comparable problems to those in the RX system developed at Stanford University (-1-). The goal of the RX system was to find clinical hypotheses from a time-oriented data base, as is that of the successor RADIX (-11-). Hypotheses finding corresponds to step 3. While in RX clinical routine data were used we deal in our system with study data of a high quality (see for the problems with routine data bases -3-,-4-).

3. THE PROSPECTIVE PATIENT REGISTRY

The registry was started in 1986, and since that time every patient cours after transplantation is monitored by a daily questionaire with the following categories of parameters:

1. diagnostically important <u>time-oriented</u> parameters like encymes (AST, ALT, AP), bilirubin, colour of bile, clotting factors,and body temperature.

2. <u>organ function scores,</u> respectively for kidney, respiratory, and circulatory function, reflecting the degree of therapeutic intervention, and coma status.

3. <u>clinical diagnoses</u> as leading hypotheses for therapy concerning hepatic or extrahepatic disorders.

4. <u>histological results</u> of liver biopsie samples classified by a transplantation pathologist and results of other clinical tests.

5. <u>therapies:</u> the type of the immunosuppression regimen, antibiotics, substitution of blood components and surgical procedures common in the posttransplant period.

These "raw data" are stored in a data base system. The data base is part of a departemental information system that is connected to the hospital information system via the MHH identification number. Lab data and other interesting data are downloaded to the PC were the system runs.

On top of this information system we built an expert system to support the investigation of "similar cases". This is a three step process. First of all there has to be done an abstracting from the raw data by clinical concept as "tendency".This concepts are represented by small procedures, that untill know incorporate heuristics, but in the near future will be evaluated by a special designed system. These procedures can be seen as complex data transformation functions in the statistical sense. We will call them here basic concepts. These basic concepts serve to build complex concepts that describe the structure of similarity of a time-oriented record. These are represented by a frame that gives the minimal structural information that have to be given so that a clinician (e.g. senior staff member) can say, they are "similar". In the liver transplant context this means that liver function and the other organ functions are to be comparable. These frames form the knowledge base. On top of this you can search in a time oriented language for similar cases, e.g. "bilirubin on high level for at least 4 days".

4. THE TRANSFORMATION (BASIC CONCEPTS)

In this first step the "raw data" are cleared from artifacts and preprocessed according to the clinical environment.

For instance body temperature is without diagnostic value when the patient's blood is ultrafiltrated to eliminate body water, and a high amount of cold fluid supplement is given. Therefore body temperature has to be marked as unreliable for those days.

Rapidly decreasing plasma encymes concurrent to a hemorrhage may mask the real pathophysiological picture. As a consequence plasma encymes are set unreliable for three days after hemorrhage to achieve a steady state condition.

In a second step specific physiological functions are scored by combining several parameters. As an example, scoring of liver synthesis capacity by plasma clotting factors has to include blood component substitution. Or, scoring the excretion capacity is a function of bile volume, color of bile, type of biliary anastomosis, and T-drain handling.

In our system we have a very rigid handling of missing values: if one parameter for scoring is missing, the score value is also missing. We think of a more flexible handling in some cases to get a better use of the informations. If there is, for instance, a strong upward trend in AST values, one missing value won't conflict with this trend.

In a third step the time-oriented parameters of the registry are condensed by trend extraction. We use a linear regression approach with heuristic placement of intersection

points. As a result the parameter curve is transformed to a
sequence of (level,trend)-pairs on a daily basis.

Partly we used the approach of VM (-5-) of different scoring
of parameter values dependent on the clinical situation
(body temperature example), but the ATN approach of Kahn
et.al. (-7-) seems to be too complex and not easily intelligible by a doctor.

We have a system under development to evaluate our heuristic
scores in the light of the clinical data.

5. MATCHING

For a given clinical situation searching for the occurrence
of the same (or "similar") situation in historical patient
courses is done by matching the patients status with entries
in a knowledge base that contains the basic concepts of the
patient. A patient's status at time t is represented by a
frame. The slot values of this frame are distinct values of
the basic concepts. The frame represent a specific kind of
"similarity" that the user (clinician) has defined before.
The match is done on a daily basis in the specified time
range.

Here an example frame:

TABLE 1

A Frame for Matching

Match on:	Concept	Basic Concept
LIVER FUNCTION	- SYNTHESIS	FV (tendency,absolute_value) substitution(much-few)
	- EXCRETION	bilirubin(tendency,level) color_of_bile (value,tend.,T-drain-stat.
	- LIVER CELL DAMAGE	transam(tendency,level LDH < 300
	- METABOLISM	ammoniak > 20 cyclo RIA/HPLC > 2
OTHER ORGAN FUNCTIONS	scores	

6. THE USER INTERFACE

The system provides the user a sreen mask corresponding to
the kind of question he wants to ask. The sreen mask consists of slot names of the corresponding frame and time parameters for the time window of the match (e.g. only day 10
to 20 after Transplantation). For a question you simply fill
in or modify the slot values for the match (you can also use
wildcard characters or include missing values) and specify a
time range. The system works in two modes:

- In the first mode you can define a patient's state of
  yourself (e.g. "What patient had normal bilirubin but
  rising transaminases?") and the system finds the days
  of patients that match this situation. You can produce
  a graphical display of interesting lab values in the
  interesting time range and inspect the other data of
  this period for each match.

- In the second mode you can choose a day in a patient's
  course and the system matches all "similar" situation.
  The same display facilities as above are avaible.

Explanation for "why"-questions is given by the slot-values
of the frame that is used for matching. "How"-questions give
the transformation rules,   the system used to compute this
value. Most interesting are "why-not"-questions: "Why not
patient E.M. between day 20 and 30 after Transplantation?"
Here the system searches for all days in the specified time
period with the fewest number of mismatched slots, and presents those. Then you can do the same displays as described
above.

7. SUMMARY AND FINAL REMARKS

We have decribed a data analyzation system that can be used
for both research and decision support purposes in the clinic. It faciliates the generation of hypotheses and structures of large data sets on time-oriented patient data, and
gives additional insights to classical statistical methods
of exploratory data analysis. It supports especially the development of clinical concepts, that easily can be viewed in
the light of the actual data.

REFERENCES

1.   Blum,R.L.: Discovery, Confirmation, and Incorporation of Causal Relationships from a Large Time-Oriented Clinical Database: The RX-Project.
Comput. Biomed. Res. 15 (1982) 164-187.

2.   Bonschek,L.I.: Are Randomized Trials Appropriate for Evaluating New Operations?
N.Engl.J.Med. 301 (1979) 44-45.

3.   Byar,D.P.: Why Data Bases Should Not Replace Randomized Clinical Trials.
Biometrics 36 (1980) 337-342.

4.   Dambrosia,J.M.,Ellenberg,J.H.: Statistical Considerations for a Medical Database.
Biometrics 36 (1980) 323-332.

5.   Fagan,L.M., Kunz,J.C., Feigenbaum,E.A., Osborn,J.J.: Extensions to the Rule-based Formalism for a Monitoring Task.
In: Buchanan,B.G., Shortliffe,E.A. (Eds.): Rule-Based Expert Systems. The MYCIN Experiments of the Stanford Heuristic Programming Project. (Addison-Wesley, Reading: 1984) 397-423.

6.   Hoaglin,D.C.,Mosteller,F.,Tukey,J.W. (Eds.): Understanding Robust and Exploratory Data Analysis.
(John Wiley, New York: 1983)

7.   Kahn,M.G., Fagan,L.M., Shortliffe,E.H.: Context-Specific Interpretation of Patients Records for a Therapy Advice System.
In: Salamon,R., Blum,B., Jorgensen,M. (Eds): Medinfo 86. Proceedings of the Fifth Conference on Medical Informatics, Washington, October 26-30, 1986. (North-Holland Publ. Co., Amsterdam: 1986) 32-36.

8.   Reichertz,P.L.: Hospital Information Systems - Past, Present, Future -. Key-note address during Medical Informatics Europe 84, 5th Congress of the European Federation for Medical Informatics, Brussels, Sept. 10-13, 1984.

9. Shortliffe,E.H.,Carlisle Scott,A.,Bischoff,M.B., Campbell,A.B., Van Melle,W., Jacobs,C.D.: ONCOCIN: An Expert System for Oncology Protocol Management. In: Proc. Seventh International Joint Conference on Artificial Intelligence. IJCAI 81. (University of British Columbia, Vancouver: 1981) 876-1015.

10. Van der Linden,W.: Pitfalls in Randomized Surgical Trials. Surgery 87 (1980) 258-262.

11. Walker,M.G., Blum,R.L.: Towards Automated Discovery from Clinical Databases: The RADIX Project. In: Salamon,R., Blum,B., Jorgensen,M. (Eds): Medinfo 86. Proceedings of the Fifth Conference on Medical Informatics, Washington, October 26-30, 1986. (North-Holland Publ. Co., Amsterdam: 1986) 32-36.

Addresses of the authors:

Günter M. Tusch
Medical School Hannover
Konstanty-Gutschow-Str. 8

D-3000 Hannover 61
============
    F.R.G

Dr. Jochen Bernauer
Medical School Hannover
Konstanty-Gutschow-Str. 8

D-3000 Hannover 61
============
    F.R.G

Prof. Dr. P.L.Reichertz
Director of the
Institute for Medical Informatics
Medical School Hannover
Konstanty-Gutschow-Str. 8

D-3000 Hannover 61
============
    F.R.G

Clinical Applications (1)

AN EXPERT SYSTEM FOR DIAGNOSIS AND THERAPY PLANNING IN PATIENTS WITH PERIPHERAL VASCULAR DISEASE

Jan L. Talmon (1, Ruud A.J. Schijven (1, Peter J.E.H.M. Kitslaar (2 and Renée Penders (3

(1 Department of Medical Informatics and Statistics, University of Limburg, Maastricht, The Netherlands

(2 Department of Surgery, University Hospital Maastricht.

(3 Vascular Laboratory, University Hospital Maastricht.

ABSTRACT

The diagnosis and therapy planning in patients with peripheral vascular disease is an area in medicine that is well suited for the development of protocols and for the application of decision support systems. In this paper the diagnostic process and the planning of therapy in these patients is outlined. Our approach to the development of a decision support system in this area will be discussed. We will conclude with the description of the first subsystem that has been developed.

INTRODUCTION

Diagnosis and therapy planning in patients with peripheral vascular disease is rather structured. Vascular disease can be suspected on the basis of the symptoms and signs of the patient. A set of diagnositic tests is used to confirm the clinical diagnosis and to quantify the severity of the disease.
First, noninvasive tests are used which are harmless to the patient. Current noninvasive vascular testing includes blood pressure determinations at various locations of the limbs, Doppler ultrasound blood flow velocity registrations at a variety of locations – combined with qualitative and quantitative analysis of the resulting waveforms – and ultrasound imaging of the vessels. These tests are often performed both at rest and under conditions of increased flow velocity (induced for instance by exercise of the limbs).
Angiography – a costly invasive diagnostic procedure which involves discomfort and some risk for the patient – is regularly needed to complete the diagnosis and to plan the therapy. The decision as to whether or not angiography should be performed is nowadays based on the results of noninvasive testing and the

possible therapeutic modalities suitable for an individual patient. The ultimate decision on surgical therapies is usually based on the results of noninvasive and invasive testing. Many nonsurgical therapies are only based on noninvasive diagnosis. In all decisions on therapy the possible outcomes of the various treatments have to be weighted with the utilities for these outcomes, both from the point of view of the patient and the physician. Computer support for the different decisions made in the vascular diagnosis and therapy may benefit from formal decision analysis techniques.

OUTLINE OF RESEARCH PROGRAM

The vascular system involved in the disease process will dictate which diagnostic tests and therapies are available. This makes it possible to develop a number of more or less independent decision aids each of which deals with specific arterial or venous regions. In our research program we start with the diagnosis and therapy of occlusive disease of the arteries of the lower extremities. In the second phase of the project we will focus on carotid artery disease. In the end, we will extend the capabilities of the system with support for the diagnosis and treatment of less common types of peripheral vascular diseases.

After a first identification of the vessels that are possibly affected, the diagnostic process is rather structured:
- A set of noninvasive diagnostic tests is to be selected
- The results of the noninvasive tests are to be interpreted
- It has to be decided which, if any, invasive tests are to be performed
- The results of the invasive tests are to be interpreted
- Finally one has to decide on the proper therapy of the patient.

Although we describe the diagnostic process as a sequential one, loops can occur. In view of the results of the instituted therapy, previously performed tests may have to be reevaluated and new testing may be required.

In order to make the development of the total decision support system manageble, we decided to start with those parts which are best defined and which involve least uncertainty and human judgement.

In the following we will describe the structure of the total system that we are going to develop. Thereafter we will describe the structure of the subsystem for the interpretation of the results of the dopplertests for assessing the condition of the arteries of the lower extremities. We will describe the methods we have used for the various functions of the subsystem. Furthermore, we will address some of the implementational issues.

GENERAL STRUCTURE OF THE SYSTEM

Since the patient data and the results of the various tests are acquired at various time instants, the decision support system has to be integrated with a database system, in which the patient data and test data are collected. For each of the different peripheral vascular systems a number of expert systems will be available, each one of them dealing with a specific task. One such task may be the selection of one or more diagnostic tests, given some global goal like assessing the severity of the disease or the possibility of surgical intervention. Other tasks are the interpretation of test results and therapy planning. The data in the database will be available to all subsystems and results of the consultations will be added to the database. Several reasoning strategies will be used, dependent on the task at hand. For example, in the therapy planning, heuristics may be used to reduce the set of possible treatments and finally, formal decision analysis may be used to determine the most appropriate one and to determine the sensitivity of the various parameters in these models. For the decision whether or not invasive testing should be performed, heuristics may be used while the search space may further be reduced by constraints that are posed by the patient data.

TESTS FOR ASSESSMENT OF OBSTRUCTIONS IN THE ARTERIES OF THE LOWER EXTREMITIES

In order to localize obstructions of the arteries of the lower extremities and assess their functional severity a number of test results are available (see for details [Pear83], for example):
- The systolic blood pressure readings taken from the upper arm, the thigh, the upper calf and the ankle while the patient is at rest. These blood pressures are determined by the Doppler method.
- The Doppler waveforms of the common femoral (groin), popliteal (knee) and ankle arteries which are visually interpreted and graded on a 7-point scale ranging from normal to highly abnormal.
- A stress test. The patient is walking on a treadmill for at most 5 minutes at a speed of 4 m/s. Before and at regular intervals after the exercise systolic blood pressures are measured at the upper arm and the ankles. The ratio of these blood pressures and the trend in this ratio provide diagnostic information. Complaints of the patient during the stress test are recorded as well as the reason for stopping the test when the patient is not able to walk for 5 minutes. All measurements are taken for the left and right leg. Basically both legs are treated separately.

ZIEKENHUIS ST. ANNADAL	HAEMODYNAMISCH ONDERZOEK
MAASTRICHT **VAATLABORATORIUM** **HEELKUNDE** **043-862272**	Patient identification

Datum onderzoek: *4 - 1o - '82* ☐ Kliniek: ☒ Polikliniek: *surgery*

Aanvragend specialist: Dopplernr.: *6451082*

Vaatoperatie(s):

VERSLAG: Both pressure-response curves are pathological. Ankle/arm-index R = 75% at rest. After exercise: 27%. Ankle/arm-index L = 75% at rest. After exercise: 39%. Patient is able to walk 3 minutes on treadmill, then experiences pain in both calfs.
The right CFA-waveform shows deminished backflow, but a normal forwardflow. The waveform in the left CFA is highpitched (stenosis). There is a fall in pulsatility between the CFA and the PA on both sides, right more than left. After the treadmilltest the waveforms in the CFA's are registered again, confirming the existence of pathology in the aorto-iliac segments.
Conclusion: pathology at the aorto-iliac level on both sides. Occluded superficial femoral artery R. Stenosis in the left common femoral artery and in the left superficial femoral artery.

DOPPLERDRUKINDICES Loopafstand: 4 km/uur: *3* min, met/~~zonder~~ pijn *calfs*

Rechts (graph, % vs RUST NA 2 4 6 8 10 12 min., Looptijd patient)

Links (graph, % vs RUST NA 2 4 6 8 10 12 min., Looptijd patient)

	RUST	NA	2	4	6	8	10	12 min.		RUST	NA	2	4	6	8	10	12 min.
Brach.	145		205	165	150	150	150	140	Brach.	145		205	165	152	150	150	140
Tib.Post. Dors.Ped.	110		55	55	55	60	70	75	Tib.Post. Dors.Ped.	110		80	90	100	100	120	115

DOPPLERREGISTRATIES	Rechts	Links	Dopplerdrukken mm Hg R	L
A. Brachialis	(Norm) / Path. + + + + + +	Norm. / Path. + + + + + +	145	
A. Femoralis Communis	(Norm) / Path. (+) + + + + +	Norm. / Path. (+) + + stenosis + + +	130	155
A. Femoralis Superficialis	Norm. / Path. + (+ +) + + +	Norm. / Path. (+) + + + + +		
A. Poplitea	Norm. / Path. + (+ +) + + +	Norm. / Path. (+) (+ +) + + +	110	110
A. Tibialis Posterior	Norm. / Path. + (+ +) + + +	Norm. / Path. + (+ +) + + +	110	110
A. Dorsalis Pedis	Norm. / Path. + (+ +) + + +	Norm. / Path. + (+ +) + + +		

+ = Geringe afplatting, Geringe backflow, Minder pulsatiel
+ + = Sterke afplatting, Geen backflow, Collaterale vulling
+ + + = Als + + doch veel meer uitgesproken

Figure 1. An example of the form for recording the results from noninvasive diagositic tests for assessment of the severity of obstructions of arteries of the lower extremities. The interpretation of these data is also shown.

In figure 1, the currently used form for recording the test results and the interpretation of the tests is shown. From the measured blood pressures in the stress test, trend plots of the blood pressure ratios are manually determined. At the top of the form, a summary of the test results is given together with an (diagnostic) interpretation of the data.

The aim of the subsystem we are currently developing is to automate the report generation process and to provide a consistent, computer-generated interpretation of the test results which may be used for subsequent higher level diagnostic and therapy planning tasks of the system.

INTERPRETATION AND REPORT GENERATION

The currently used reports are already quite structured: they summarize the results of the stress test, they describe the abnormalities found in the doppler waveforms and finally give an interpretation.

Rather standardized texts are used in these reports. Hence we decided to use the Augmented Transition Network (ATN) approach, similar to the one used in ATTENDING - [Mill83] - for text generation. Figure 2 shows a small part of the ATNs that were designed to cover all possible reports.

Each arc of an ATN represents either some text or another ATN, which will be displayed when a set of associated criteria is met. Parallel arcs, which leave a node, are alternatives. The one for which the criteria are met is traversed. Then, the selection at the node that is reached is made. The criteria may either refer directly to the results of the diagnostic tests or refer to more global concepts that are derived form the test results. For example, when the text "An abnormal response to the stress test is found in both legs" has to be displayed, one can either define the criteria for this statement directly in the ATN representation or one can introduce a concept like BOTH.LEGS.INVOLVED, which in turn may be derived from rules like "if LEFT.LEG.INVOLVED and RIGHT.-LEG.INVOLVED then BOTH.LEGS.INVOLVED"

The latter approach is prefered because the report generation process will in general be based on more global concepts. The rules and criteria that describe how these global concepts relate to the test results comprise the domain knowledge. This knowledge is stored in a separate rule base.

The inference process is a form of backward chaining. The ATNs are traversed. When the facts, mentioned in the criteria along the arcs are not found, the knowledge base is consulted in order to possibly derive that fact in a goal-driven manner.

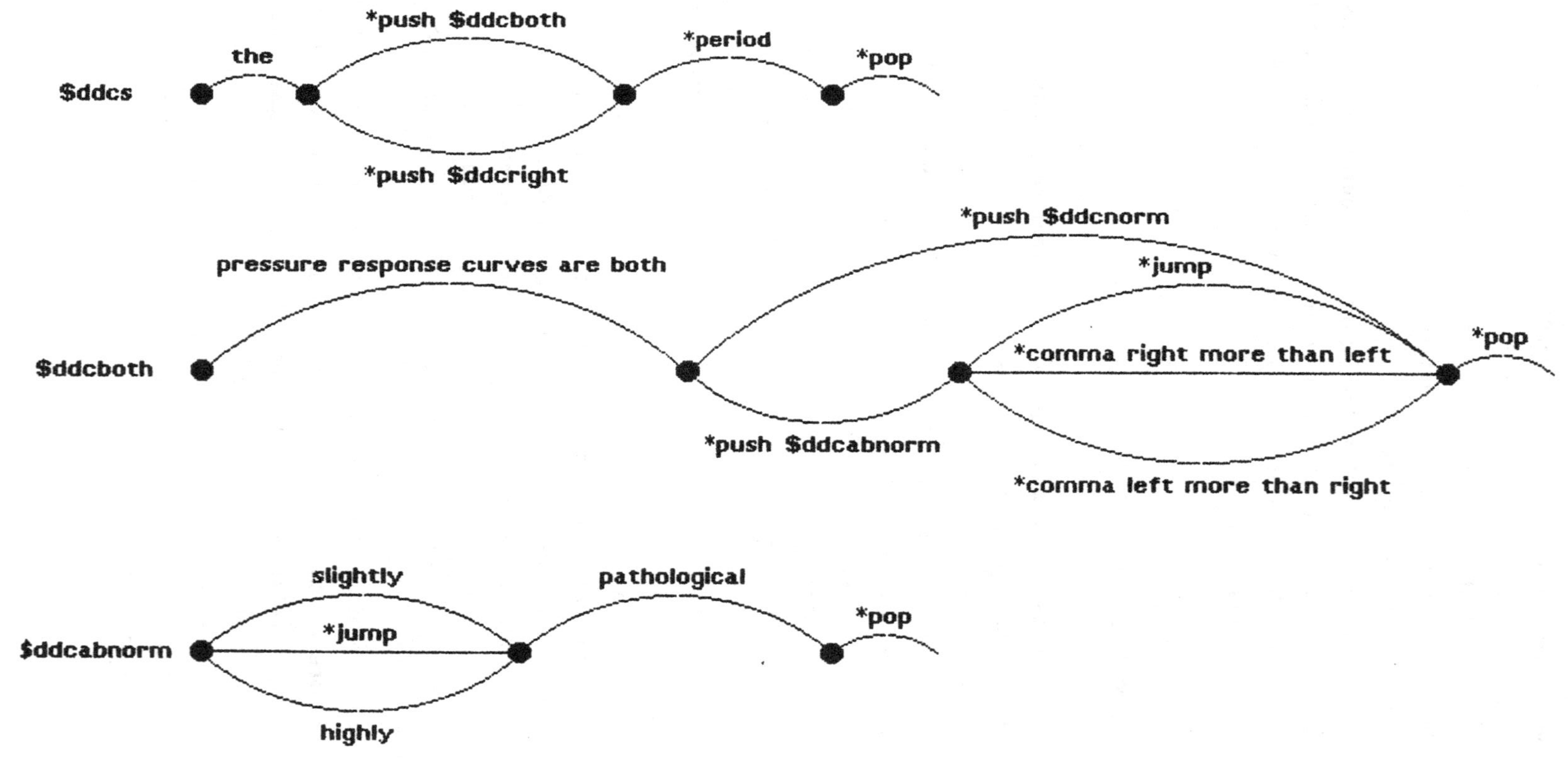

Figure 2. Some of the ATNs used in our report generation. The text generation starts with arc $ddcs and prints "the". Next a selection is made of the two arcs, dependent on which criteria are met. *push indicates that the text generation has to continue with another ATN, like $ddcboth. *period and *comma indicate which punctuation is to be used. *jump indicates that no text is associated with that arc, while *pop returns to a higher level ATN.

KNOWLEDGE ACQUISITION

The knowledge to be stored in the rule base is collected in the following way. From a large set of previously performed tests, the forms are still available. Test results and interpretations were stored in a database for this project. Decision rules were induced from examples, using a previously developed system, [Talm86]. The algorithm we use is similar to the ID3-algorithm described by Quinlan (Quin86). A basic difference is that our algorithm can also handle noisy data. This makes it possible to identify possibly wrong interpretations. For the system, described in here, we used the algorithm to derive criteria for the interpretation of the test results. Inconsistencies, resulting in an incomplete separation of the cases into the different interpretations, were discussed with clinicians and expert technicians. Previous interpretations were modified when obvious misinterpretations were made. Several optimization runs were performed. Furthermore, we used the obtained decision tree to find conceptual descriptions of the interpretations in terms of constraints on the test results (see [Lee86]).
These conceptual descriptions were used to discuss the validity of the derived criteria. When necessary, examples from the training set were used to illustrate the concepts.
This process resulted in a rule base for the interpretation of the test results which is currently tested in clinical practice.

IMPLEMENTATION

The complete system has to be operational in a hospital environment.
Data from patient contacts will become available at other places than the test results. Therapy planning etc will also be done at different places. Hence, the system has to run on a multi-user system, possibly on distributed hardware. We opt for having decision support systems of the type described in this paper on (powerfull) PCs. This poses some restrictions on the implementation, specifically with respect to the user interface. The subsystem described in this paper has been implemented on an Olivetti M24 and is written in muLISP. Currently, the underlying database system is not yet implemented and the subsystem can operate as a stand alone system.
A database of the test results is maintained on the system. Data entry, for the time being, is very much structured. The user is prompted for the test results. Further developments with respect to the user interface are necessary to integrate the system into routine clinical practice.

SUMMARY AND CONCLUSION

We described a research program for the development of a decision support system for the diagnosis and treatment planning for patients with peripheral vascular disease. The structure of the domain was outlined. We reported in detail on the subsystem for the interpretation of the results of tests for the non-invasive assessement of the severity of obstructions of the arteries of the lower extremities. The tasks of the subsystem were described and the knowledge representation and inference procedure discussed in the context of report generation.

The approach we have taken enables us to develop a rather complex decision support system in an incremental way. We can focus on those parts that are most useful from a clinical point of view either because they will make the decision process more objective or because they will contribute significantly to the efficiency of the diagnostic process.

REFERENCES

[Lee86]   Lee, W.D. and S.R. Ray: Probabilistic rule generator. Report
          UIUCDCS-R-86-1263. Dept. of Comp. Science, University of Illinois,
          Urbana-Campaign, Ill, USA.

[Mill83] Miller, P.L.: ATTENDING: Critiquing a physician's management plan.
          IEEE Trans. PAMI-5, 449-461, 1983.

[Pear83] Noninvasive vascular diagnositc testing. Curr. Probl. Surg., 20, 469-
          491, 1983.

[Quin86] Quinlan, J.R.: Induction of Decision Trees. Machine Learning, 1, 81-
          106, 1986.

[Talm86] Talmon, J.L.: A multiclass nonparametric partitioning algorithm.
          Pattern Recognition Letters, 4, 31-38, 1986.

AN EXPERT SYSTEM FOR THE CLASSIFICATION OF DIZZINESS AND VERTIGO.

R. Schmid (1), P. Zanocco (2), A. Buizza (1), G. Magenes (1), M. Manfrin (2), E. Mira (2).
(1) Dipartimento di Informatica e Sistemistica and (2) Clinica Otorinolaringoiatrica, University of Pavia, Pavia, Italy.

The syndromes characterized by symptoms of vertigo and/or dizziness represent a quite complex field whose physiopathogenetic aspects are not yet completely clear. They can involve the peripheral vestibular system, the vestibulo-cerebellar pathways, and the central nervous system (CNS). A knowledge based consultation system, or expert system (ES), could have useful applications in this field, both as clinical tool for differential diagnosis of vertiginous syndromes and as educational tool in the Schools of Medicine, in particular of Otoneurology. The construction of such an ES would be an opportunity to organize the available knowledge in a rational and clear framework, and through its use the effectiveness of the current diagnostic methods and procedures could be objectively tested. In particular, it seems worth testing how far one can go with differential diagnosis of vertigo when only information concerning patient history is considered. Nowadays, the importance of history taking is often underestimated and neglected in favour of more complicated and expensive instrumental tests. In contrast, our working hypothesis is that this phase of the diagnostic procedure makes available precious and unique information, provided it is conducted and exploited in expert way, and that, in many cases, a quite complete and reasonably definite diagnosis can be based on history taking.

The goal of the present research is to build a consultation model to be used in a clinical and educational environment as a diagnostic aid for vestibular disorders. As a first step, we restricted our attention to pathologies which present vertigo or dizziness as a major symptom. The ES so far developed (VERTIGO) is able to propose differential diagnosis based on data arising from patient history. Its main features will be discussed in the following sections.

METHODS

VERTIGO has been developed by using the Rutgers University EXPERT shell for knowledge representation [4-6] that has already been used in a number of clinical consultation models [2,3,5].

An EXPERT consultation model is made of findings, hypotheses, and decision rules. The latter are production rules that infer hypotheses (e.g. diagnostic conclusions) from the set of findings (e.g. patient data). They can relate findings to hypotheses (FH rules) or hypotheses to each other (HH rules). The strenght of each conclusion is quantified by figures (confidence weights) lying in the interval -1 (absolute disproof) to +1 (absolute certainty). The hypotheses can be organized in a taxonomic structure. "Intermediate Hypotheses" and "Treatments" can also be defined. The findings are collected through a system driven questionnaire.

EXPERT, written in FORTRAN, runs under VMS on the VAX 8500 of the Computer Centre of the University of Pavia.

KNOWLEDGE BASE

It is important to define the limits and the field of application of VERTIGO in its present form.

First of all, VERTIGO has been conceived to provide classification
of vertigo and dizziness and not to make general otoneurological
diagnosis. Consequently it deals only with patients who complain at
present, or complained in the past, of vertigo and/or disequilibrium
or spatial disorientation related, in some way, to vestibular
disorders. Dizziness due to mental or ophthalmological troubles,
once individuated is discarded without further analysis.
The second limitation is that attention has been mainly paid to the
differential diagnosis of vertigo due to peripheral vestibular
disorders. All forms of vertigo of central origin are clustered in
one diagnostic category that will be split up later.
The third and more important limitation is that only data derived
from patient history are considered as findings. Objective signs and
results of instrumental tests will be taken into account in a
further phase of the project development. The rationale for this
choice is not only to simplify the construction of the system by a
step-by-step strategy. Rather, it corresponds to the precise goal of
re-evaluating the importance and the uniqueness of history taking in
the clinical practice. This attitude has recently been stressed by
many orl specialists just in relation to the differential diagnosis
of vertigo [1]. In developing VERTIGO, a special care has therefore
been devoted to this aspect in order to obtain a tool for verifying
the accuracy and the reliability of the diagnostic conclusions that
can be reached through a skillful examination of patient history.
This voluntary limitation of the horizon of VERTIGO gives a further
justification to the other limits of the system that have been
previously outlined. As an example, detailed classification of the
forms due to CNS troubles necessarily requires the results of
instrumental tests.
Table 1 gives the hypotheses considered by VERTIGO, that is, the set
of vertiginous syndromes it can classify and their taxonomy.
Hypotheses are clustered in two main groups: "Vertigo" and
"Dizziness". The former contains pathological situations in which
the main revealing symptom is a sensation of rotation and/or
translation concerning either the patient himself (self-motion

sensation) or his environment (object-motion sensation). The latter refers to situations mainly characterized by disequilibrium and spatial disorientation that can still be referred to vestibular disorders. The subgroup "Non-Vestibular Dizziness" in the group "Dizziness" contains the forms due to mental or ophthalmological rather than vestibular problems. As previously reported, once identified, they lead to the immediate end of the diagnostic session.

In turns, the group "Vertigo" includes two subgroups, "Paroxysmal Vertigo" and "Sudden Loss of Vestibular Function". In both, forms of vertigo different in aethiology are put together according to their common morphology. Namely, all the forms of paroxysmal vertigo present repeated and very short attacks lasting several seconds or a few minutes. Vertigo due to sudden loss of vestibular function is an isolated and very heavy attack lasting more than a day.

The remaining vertiginous syndromes present more individual morphologies, so they are not clustered.

Since the taxonomy is structured according to morphology, the same aethiology may appear as the origin of different forms of vertigo. For instance, the vascular troubles are considered in relation to any kind of vertigo and dizziness since they are one of the most common causes of vestibular disorders. Similarly, CNS troubles are considered as a possible cause of both vertigo and dizziness.

Findings, that is patient's history data, are collected through a questionnaire. After unspecific questions concerning the age, the sex and the general condition of the patient, the key-question is asked in order to know whether the patient is really complaining of vertigo. A positive answer triggers a set of questions aimed to define the characteristics of the vertigo attacks (age of the symptoms, modality of appearance, frequency of repetition, duration and strength, concomitant nausea, existence of precipitating factors).

If the key-question is given a negative answer, the system switches to a different set of questions aimed to know whether the patient complains of other symptoms (imbalance, spatial disorientation, blurred vision, sequelae of past vertigo attacks) that could be referred to vestibular disorders. If only vision troubles (diplopia, scotomata, phosphenes, hemianopsia) or mental troubles (loss of consciousness, mental confusion, loss of concentration) are reported, the questionnaire stops and the diagnosis of "Non-Vestibular Dizziness" is proposed.

Except for this last case, the questionnaire goes on seeking for the presence of aethiopathogenetic factors (cardio-vascular diseases, cranio-cervical traumas, intoxication from ototoxic drugs or chemicals,...) or concomitant signs (hearing loss, cranial nerve troubles,...).

In this questionnaire, different symptoms elicit different questions so that only data relevant to the classification of the troubles the patient is complaining of are acquired. For instance, since cardio-vascular disorders can be the origin of any kind of vertigo and dizziness, every patient is questioned about the condition of his cardio-vascular system, regardless of the symptoms that drove him to undergo a medical examination. On the contrary, only patients who complain of dizziness are questioned about possible intoxication from ototoxic drugs, as it is known that dizziness, rather than vertigo, is the usual manifestation of such an intoxication. In this way not pertinent or frankly stupid questions are avoided, while the system emulates the questioning logic of an expert clinician. As a matter of fact, what he usually does is to seek for the information needed to confirm or to disprove a diagnostic suspect formulated on the ground of a few key-symptoms and updated as new data are acquired. So, he adapts his questioning strategy to the available information.

At present the questionnaire contains about 40 questions concerning a total of about 90 findings. Depending on question selection, no more than 15 to 25 questions are actually asked during a typical diagnostic session. Since only patient history is taken into account, many data are based on subjective judgement of either the patient or the physician. This introduces a degree of uncertainty that we tried to reduce by careful formulation of the questions.

Owing to the features of the shell EXPERT, the knowledge base of VERTIGO is formalized by means of production rules. Intermediate Hypotheses are evaluated from Findings through FH rules. They represent typical patterns of findings describing morphological characteristics of the vertigo (e.g. "Severe and Long Lasting Vertigo Attack"), association of vertiginous symptoms with extra-vestibular pathologies or signs (e.g. "Cochlear Signs Associated with the Vertigo Attacks"), predisposing factors of the patient (e.g. "Subject Physically Disposed to Vertebro-Basilar Insufficiency"), etc. .

The final diagnostic conclusions and their weights are evaluated by HH rules that link Intermediate Hypotheses to each other and to Findings. According to the criteria adopted in the Taxonomy of VERTIGO, Intermediate Hypotheses dealing with different morphologies of vertigo are used to trigger different subsets of HH rules. Each subset contains the rules relevant to the vertigos that present a given morphology. However, when pathognomonic signs exist, they usually prevail over morphology and are used as triggers.

At present the knowledge base of VERTIGO contains about 350 rules. Special care has been devoted to manage the uncertainty related with the qualitative nature of most of the findings.

RESULTS AND CONCLUSIONS

In its present form, VERTIGO has been tested on about 150 cases taken from every day clinical practice. A few very rare forms (e.g. Vertigo due to Hepilepsy) were never encountered and could not be submitted to the system. Our main goal in this phase was to make the model to reproduce as closely as possible the reasoning of the orl specialists that participated in its development. At present, the level of disagreement between the diagnostic conclusions of VERTIGO and those independently reached by its constructors is comparable with that existing among the diagnostic conclusions of different human experts. Then, a more thorough validation on a larger and more complete data base is being undertaken with the help of specialists who did not collaborate in the project. This validation, the use of data from objective examination and instrumental tests as further findings, the enlargement of the knowledge base with better definition of CNS pathologies, and the implementation of a PC version of the system will be the lines for the future development of VERTIGO.

Finally, a few remarks are worth-while about the use of the shell EXPERT in this particular context.

The possibility of structuring the Hypotheses in a taxonomy is a useful feature of EXPERT since the diagnostic conclusions can be organized in a clear structure and predecessor-successor relationships can be defined. In the present case, however, organizing the data in a net would have been even more useful since not only morphological (as we did) but also aethiological relationships could have been represented.

The more intriguing problem we had to face when developing VERTIGO was the control of the questionnaire, as the capability was needed to select the questions according to diagnostic suspects elicited by the available information. Since in EXPERT the control of the reasoning process is data-driven, Hypotheses cannot be used to guide the questioning strategy. Flexibility of the questionnaire can be obtained only by using rules relating Findings to Findings (FF rules). They allow to infer the value of new findings from that of other already known findings and to set a finding as "unknown" when information is acquired that makes the knowledge of that finding useless or the relevant question not pertinent. Then, all the questions concerning findings whose value has already been set by FF rules are skipped.

The structure of FF rules does not permit to define logical operations among the findings in the antecedent of a rule. The OR operation may be easily simulated by a set of rules with different antecedents and one consequent. However, the control of our questionnaire required both OR and AND operators. We overcame the intrinsic limitation of FF rules and implemented AND operators by means of "OR sets" through the well known De Morgan's formula. To do that, we had to define appropriate "dummy findings", that is, findings whose value is always set by FF rules and never asked to the user. With respect to the other findings, "dummy findings" are used only to control the sequence of the questions and do not contain information to be used to diagnostic purposes.

Probably, the control of the questionnaire would have been simpler in a backward-chaining, or hypotheses-driven, environment. Here, diagnostic suspects, when elicited, directly guide the questionnaire looking for the specific piece of information needed for their own proof. To emulate this behavior, EXPERT requires that part of the knowledge contained in FH and HH rules be repeated, with more difficulties, in FF rules. So, EXPERT seems more suitable to manage problems requiring a quite stiff questionnaire with simple questioning strategy rather than problems where a flexible and adaptable questioning approach should be adopted.

Apart from these few limits, which, perhaps, are mainly dependent on the specific application we are dealing with, EXPERT is a quite easy to use and performant shell. It gives well structured consultation models and shows a good level of interactivity.

REFERENCES

[1] Collard M., Conraux C., Freyss G., Haguenauer J.P., Legent F., Perrin C., Sans A., Van Cauwenberge P. (1986) MDV1 - Manuel Diagnostique des Vertiges. Villeurbanne: Duphar et Cie.
[2] Kastner J.K., Dawson C.R., Weiss S., Kern K., Kulikowski C. (1984) An expert consultation system for frontline health workers in primary eye care. J.Med.Syst. 8: 389
[3] Quaglini S., Stefanelli M., Barosi G., Berzuini A. (1986) ANEMIA: an expert consultation system. Comp.Biomed.Res. 19: 13-27.
[4] Weiss S., Kulikowski C. (1979) EXPERT: A System for Developing Consultation Models. CBM-TR-95, Department of Computer Science, Rutgers University.

[5] Weiss S., Kulikowski C., Safir A. (1978) Glaucoma consultation
by computer. Comp.Biol.Med. 8: 25-40.
[6] Weiss S., Kern K., Kulikowski C., Uschold M. (1984) A Guide to
the Use of the EXPERT Consultation System. CBM-TR-94, Department
of Computer Science, Rutgers University.

```
**HYPOTHESES
*TAXONOMY
SV        Vertigo
PAR           .Paroxysmal Vertigo
INFA              ..Benign Paroxysmal Vertigo of Childhood
POSI              ..Benign Paroxysmal Positional Vertigo
INVB              ..Paroxysmal Vertigo due to Vascular Disorder
ORTO              ..Orthostatic Vertigo
CERV              ..Cervical Vertigo
HEPI              ..Hepileptic Vertigo
FIST              ..Paroxysmal Vertigo due to Perilymphatic Fistula
GCV           .Sudden Loss of Vestibular Function (Severe Vertigo Attack)
GCNV              ..Severe Vertigo Attack due to Vestibular Neuritis
GCN8              ..Severe Vertigo Attack due to VIII c.n. Neuritis
GCIV              ..Severe Vertigo Attack due to Vascular Disorder
GCHZ              ..Severe Vertigo Attack due to Herpes Zoster Oticus
GCFR              ..Severe Vertigo Attack due to Temporal Bone Fracture
GCLP              ..Severe Vertigo Attack due to Suppurative Labyrinthitis
MEN           .Meniere's Disease Vertigo
SCG           .Cogan's Syndrome Vertigo
VEM           .Vertigo due to Migraine
VVP           .Vertigo due to Vascular Disorder
VSC           .Vertigo due to Possible CNS Troubles
NRA           .Atypical Vertigo due to Cerebellopontine Angle Tumor
VHZ           .Atypical Vertigo due to Herpes Zoster Oticus
VTR           .Atypical Vertigo due to Temporal Bone Trauma
SS        Dizziness
LAS           .Dizziness due to Serous Labyrinthitis
NRM           .Dizziness due to Cerebellopontine Angle Tumor
DVD           .Dizziness due to Vascular Disorder
OTF           .Dizziness due to Ototoxic Drugs
OTI           .Dizziness due to Ototoxic Chemicals
ICF           .Dizziness due to Incomplete Vestibular Compensation after SLVF
MFA           .Dizziness due to Old Aged Meniere's Disease
SCS           .Dizziness due to Temporomandibular Joint (Costen's) Syndrome
SSC           .Dizziness due to Possible CNS Troubles
NVD           .Non-Vestibular Dizziness
PSNT              ..Dizziness originating from Mental Troubles
OFTT              ..Dizziness originating from Ophthalmological Troubles
```

Table 1

Work supported by CNR and by HUSPI Project

THE SENEX SYSTEM : A MICROCOMPUTER-BASED EXPERT SYSTEM BUILT BY ONCOLOGISTS FOR BREAST CANCER MANAGEMENT

Jean-Louis RENAUD-SALIS, Françoise BONICHON, Michel DURAND,
Antoine AVRIL, Claude LAGARDE,
J.P. SERRE, P. MENDIBOURE

Fondation Bergonié, 180, rue de Saint-Genès, 33076 Bordeaux Cédex
(France)

SUMMARY

The SENEX project started in the beginning of 1985 at the Fondation Bergonié, a comprehensive cancer center located in Bordeaux, FRANCE.

The major general goal of this project is to demonstrate the feasability of cancer experts developing efficient consultation systems that can run on affordable microcomputers.

This could help to manage clinical trials and disseminate state of the art cancer treatment protocols issued from cancer clinical research.

The first specific goal of the project was to develop a prototype system applied to breast cancer treatment (the SENEX system) with the aid of a knowledge engineering package (Personal Consultant Plus, Texas Instruments).

This paper reports on the successive phases of the knowledge base development, the current system status and the problems encountered. The SENEX system has reached the stage of a research prototype. It embeds approximately 400 production rules attached to 40 frames and performs at the expert-level for in-protocols patients. Its control structures are modeled from the ONCOCIN system.

Formal evaluation will take place within this year.

Current works and future projects include the integration of cancer consultation systems with a temporal clinical database management system (MEDLOG) and a statistical knowledge base. Altogether, these programs are intended to constitute a complete cancer clinical research system that can run on inexpensive microcomputers.

THE SENEX SYSTEM : A MICROCOMPUTER-BASED EXPERT SYSTEM BUILT BY ONCOLOGISTS FOR BREAST CANCER MANAGEMENT

Jean-Louis RENAUD-SALIS, Françoise BONICHON, Michel DURAND,
Antoine AVRIL, Claude LAGARDE,
J.P. SERRE, P. MENDIBOURE

Fondation Bergonié, 180, rue de Saint-Genès, 33076 Bordeaux Cédex
(France)

1. Introduction

As clinical research in oncology progresses, it provides the community of physicians involved in the management of cancer patients with a growing and rapidly renewed body of empirical knowledge. This continually widens the gap between what a physician should know about cancer and what can be retained and utilized.

Furthermore, cancer treatment protocols are increasingly complex and difficult to manage. As these protocols serve to standardize and coordinate the treatment, they are essential for the management and the analysis of clinical trials. Moreover, they constitute when validated a useful way for disseminating the state of the art knowledge (1,2). Whereas it has been suggested that wide application of cancer treatment protocols could dramatically reduce the overall mortality from cancer, recent studies have emphasized the difficulties to undergo and manage large clinical trials and to disseminate widely the knowledge acquired from these trials (3).

The evaluation of the ONCOCIN system (a knowledge-based system for the management of cancer chemotherapy protocols developed at Stanford University) has demonstrated that expert-systems could have a major impact on the collection of clinical data, and can provide expert-level advices in the management of cancer patients (4,5).

It can therefore be assumed that the dissemination of such computer-assisted consultation systems can help to overcome the difficulties encountered in the management of clinical trials, to spread state of the art cancer protocols and to facilitate their application.

Due to the absence of high-level development tools and the exigence for expert-systems to run on large and costly computers, the human and financial costs involved in the development of such systems have until now confined their use in research laboratories. Fortunately, things have changed in the past few years and tools exist today that are claimed to allow medical experts with little experience in computers and in medical knowledge engineering to teach computers what they know and to develop effective consultation systems that can run on inexpensive microcomputers. This convergence between ever-growing needs in the cancer domain and the fast-increasing cost-effectiveness ratio of the microcomputers and software technologies led us to start upon the SENEX project at the Fondation Bergonié at the beginning of 1985.

The major general goal of this project was to demonstrate the feasability of cancer experts developing efficient cancer consultation systems with the aid of expert systems development tools. This paper relates our experience in the development of a breast cancer knowledge based consultation system intended for helping physicians in the management of clinical trials and for disseminating the knowledge acquired from these trials.

2. Overview of the SENEX project

2.1. The overall objectives of the project are :

. to evaluate whether efficient knowledge based consultation systems that can run on microcomputers can be developed by cancer experts with the aid of expert-systems development tools.

. to evaluate the strengths and weaknesses of the development tools and to establish effective relationship with researchers and companies involved in the development of I.A. tools in order to improve these tools and to adapt them to the cancer domain problems.

. to establish a general methodology for building cancer protocols management systems.

. to integrate knowledge-based consultation systems with a medical database management system and statistical programs and to realize eventually a clinical research workstation system.

2.2. Specific goals

Our first specific goal was to build a consultation system intended for advising physicians in general hospitals in the management of breast cancer treatment protocols.

We chose breast cancers as our first application domain because of the model status of breast cancer in oncology, the complexity of breast cancer protocols, and because a significant number of breast cancer patients are usually taken into care by several institutions during their treatment course. Furthermore, as the Fondation Bergonié is a leading cancer center in breast cancer management, breast cancer experts were available and highly motivated.

The following goals have been assigned to the SENEX system :

. to give advices on breast cancer management based upon the formal knowledge contained in protocols and some experiential or judgmental knowledge captured from the experts.

. to improve data collection, management and analysis of clinical trials.

. to generate automatically and print reports related to patient management (summary sheets, progress notes, prescriptions...)

. to be fast and congenial enough to be acceptable and usable by physicians with a minimal training in computer use.

3. Overview of the development process

After the completion of a feasability study aimed to choose the hardware and software resources to be used, we selected the Personal Consultant development tool (Texas Instruments) for building our first prototype. This tool is a microcomputer adaptation of the EMYCIN system which provides in its current Personal Consultant Plus version (PC Plus) most of the desirable features for the development of a mid-sized medical consultation system (production rules, frames, meta-rules, procedural capabilities, communications with external programs, user-defined LISP functions...). It runs on IBM and compatible microcomputers.

3.1. Conceptualization and Formalization phases

The main body of knowledge comes from the breast cancer treatment written protocols established at the Fondation Bergonié since 1984. The successive tasks of conceptualization and formalization of the knowledge have been greatly facilitated by the existence of this readily available formalized core of knowledge, and by the participation of an experienced physician specially trained in medical knowledge engineering.

Moreover, the availability of numerous papers from the Stanford group reporting on their experience in building the ONCOCIN system greatly furthered these stages as well as the following ones.

3.2. Implementation phase

After a training period of a few weeks, we were able to use the PC Plus tool on a routine basis and to build a Mark I system. This trial and errors initiation provided us with more confidence in the system capabilities and allowed us to gain some insights into the system reasoning process. We did not encounter too much difficulties in writing rules with the aid of the abbreviated rule language. Most of the problems were related to the building of control structures.

The first demonstration prototype took approximately 350 man/hours to be developed and its performances were encouraging (6).

4. Knowledge representation in SENEX

4.1. Overview of the cancer treatment domain

Written cancer protocols are hierarchically structured. The successive steps of the general problem solving strategy are well established and consistently organized through the various protocols. The protocols consist of a finite set of clinical situations corresponding to a finite set of therapeutic solutions which can be enumerated in most of the cases.

Most of the knowledge embedded in formal protocols is categorical.

These characteristics of the general domain knowledge and procedures correspond to classification problems and categorical reasoning, apart from unusual cases which may require some probabilistic reasoning.

On the other hand, cancer knowledge is essentially empirical, as the actual mechanisms of disease processes and treatment processes are poorly understood. This lack of significant fundamental knowledge about the cancer disease and its treatment explains the need for clinical experiments, and the wide use of heuristic rules in cancer patients management.

These features make the expert-systems approach very suitable to the cancer domain.

4.2. General problem solving strategy

The successive steps of the general problem solving strategy can be briefly described as follows :

General characteristics of the patient are first collected, and informations describing the tumor and its extension are required. Results of additional tests are then asked from the user, and inferences are made that conclude the stage of the disease, and the prognostic category. Heuristic rules then establish a relation between the prognostic category and the desirable treatment plan. Next refinement steps occur until the detailed prescription is established. Chemotherapies are the most complex treatments : they are made of one or several successive chemotherapy protocols and each protocol comprises one or more drugs. Drugs are prescribed according to detailed specifications related to the scheduling and the procedures of administration of the drugs and instructions to modify the doses of drugs in case of toxicity.

4.3. Domain-specific knowledge representation

Most of the domain-specific knowledge has been embedded in rules and frames.

. Production rules are used to represent the inferential knowledge. They establish the stage of the disease, the prognosis, the treatment plan, etc... They are attached to specific frames.

An example of a specific rule attached to the treatment plan frame is : if the stage of disease is advanced, loco-regional and the patient is to be included in a research protocol, then apply protocol no. 3 and chemotherapy will be given as first treatment.

Goal-driven rules and data-driven rules are intermingled in the SENEX system. The latter and active-valued parameters are used to provide values to the parameters describing the detailed specifications of the treatment protocols, and to execute specific procedures such as body surface area and drug doses calculations.

The linking of rules and procedures to specific frames explicitly defines the contexts, within each rule or set of rules is applicable.

. <u>Frames</u>

Frames are organized according to hierarchical and causal links. They represent concepts and entities such as patients general characteristics, diagnostic workup, treatment plan, treatment protocols, chemotherapies, drugs, etc...

The frames for each of the chemotherapies and drugs contain slots to which are attached procedures for the calculation of these variables, or slots for the respective doses and routes of administration (Fig 1).

. <u>Management of uncertainty and incomplete knowledge</u>

As the current knowledge base contains essentially the categorical knowledge embedded in formal protocols, we have not yet used certainty factors. Default values are used to provide values to some parameters such as drug toxicity, metastatic status, routes of administration, etc...

. <u>Temporal links</u>

The SENEX system does not currently manage temporal relations between parameters. This problem will be addressed at a later stage when the interface between PC Plus and a temporal database management system have been completed.

4.3. <u>Control structures</u>

Rules and frames have been used for the representation of control procedures.

. <u>Rules</u> are used for specifying general and local context dependent procedures as in the following examples :

If the goal of the consultation is to establish the initial treatment,
and the diagnostic workup hase been done,
and the treatment plan has been established,
and the patient accepts the proposed treatment,
then, apply the proposed treatment.

Or for a more specific context dependent procedure :
if the patient has undergone a lumpectomy and an axillary dissection,
and information has been gathered about the histological findings of the specimen,
and information has been gathered about the hormonal dependence of the breast tumor,
then, the post-operative workup is complete.

. <u>Data blocks, Control blocks</u>

As in the ONCOCIN system (7) data and control blocks are used to perform specific tasks and to execute successive steps in a consistent order. They can request together blocks of logically related parameters.

Control blocks can be menu driven and are invoked whenever the user specifies the task SENEX is to perform or whenever the reasoning process needs it.

. <u>The frame structure</u> (Fig 2) provides the overall control for the problem solving process and allows to focus the reasoning of the system. The structural and procedural knowledge of the chemotherapy protocols is represented in Fig 1.

5. <u>Current status and performances of the SENEX system</u>

The SENEX system has reached the stage of a research prototype. It provides the following features :

. Establishment of the stage and prognosis of breast cancer patients.

. Recommendations on the appropriate treatment plan and management of protocols according to the individual characteristics of the patients.

. Establishment of detailed prescriptions.

. Printing of summary sheets, progress notes and prescriptions.

. Simple explanations about its line of reasoning and the inferences made.

The system performances have been evaluated informally, after a review of 120 patients charts, by comparing the system advices to the actual decisions made by physicians.

The recommendations provided by the system are in agreement with the actual decisions in more than 90 % of the cases. Most of the disagreements occured for patient not included in protocols and representing unusual cases (Table 1).

Table 1

INITIAL EVALUATION OF THE SENEX SYSTEM PERFORMANCES

| | Agreement | Disagreement | | No advice provided by |
		Acceptable	Inacceptable	SENEX
Routine Protocols	44/50	3/50	/	3/50
Research Protocols	48/50	2/50	/	/
Out of Protocols	15/20	3/20	/	2/20
TOTAL	107/120	8/120	/	5/120

The current system consists of about 400 rules and 40 frames.

The user interface provided with the PC Plus development tool is congenial. The average time for a consultation of the system is approximately four minutes.

6. System weaknesses, Current works and future projects

6.1. Integration of the SENEX system to a temporal database management system

One of the main weaknesses of SENEX is its inability ot interact with a database management system. This requires the user to reenter past informations on patient and treatment history at every new encounter. This prevents also the system to be effectively used for the management of clinical trials.

An interface between PC Plus and a time oriented medical database management system (MEDLOG - Information Analysis Corporation) has been developed in collaboration with the

French Company distributing the latter. This interface consists of LISP functions and external routines (written in Assembly and C language) that enable SENEX rules to get values from variables already entered in the patient database (historical data, current treatment, cumulative doses of drugs, etc...).

This interface provides a useful mechanism for assessing trends and searching data according to temporal criteria. In the other way, this interface allows to update the patient database with the new data acquired from every consultation of the SENEX system. As the MEDLOG system provides a complete set of biostatistical programs, the integration of the PC Plus and MEDLOG systems will eventually offer a full set of functions covering most of the needs of clinical researchers.
The SENEX knowledge base is currently being reimplemented using these new features and this work is scheduled to be completed within the next six months.

6.2. User interface

Although the PC Plus user interface is reasonably congenial, it does not allow the user to take control over the system and to volunteer data or suggest hypotheses at any time during the consultation process. This lack of flexibility is inherent to the design of EMYCIN like systems such as PC Plus and could harm the system acceptability in actual clinical practice.

However, this constraint can be useful in the context of clinical trials or good clinical practice analysis as it will oblige physicians to follow a consistent path in data collection.

For routine uses, we envisage as a solution to implement future knowledge base on more flexible systems such as PROPS 2 (a PROLOG-based system developed at the Imperial Cancer Research Fund Laboratory (8)) or NEXPERT (Neuron Data Corporation).

Another weakness of the current SENEX system is its inability to display data tables allowing the user to expedite the data acquisition process for logically related parameters.

This feature will be implemented as a standard feature in the forthcoming PC + 3.0 version.

6.3. Knowledge base insufficiencies

The current SENEX knowledge base lacks judgmental or experiential knowledge for dealing with out-of-protocols and unusual cases. We are currently working with our breast cancer experts to capture this essential knowledge. This will lead to the use of uncertain reasoning and deeper knowledge representation.

6.4. Formal evaluation of the SENEX system

As our major goal is to allow cancer experts to develop consultation systems that can have an impact on the quality of clinical trials and on the dissemination of state of the art cancer knowledge, our evaluation program is intended to assess the performances of the system, as well as its mechanisms of reasoning, its acceptability and its cost-effectiveness. We shall follow the methodology already established by the Stanford Knowledge Systems Laboratory (4,5). This evaluation and refinement process will take place within the year following the initial structured evaluation of the system performances by the breast cancer experts at Fondation Bergonié.

Three groups of physicians involved in cancer patients management in general hospitals have already accepted to test the system in their actual practice.

6.5. Future developments

Further development projects include the adjunction of a knowledge based statistical system intended for advising physicians in the analysis of medical databases, and the development of new knowledge based systems in other cancer domains.

Discussion and Conclusions

Using a commercially available knowledge based system development tool, we have been able to build an efficient, although still limited, system for advising physicians in the management of breast cancer patients. Our first general goal, which was to demonstrate the feasability of cancer experts developing knowledge based consultation systems with the aid of expert systems development tools has been achieved.

The SENEX system development has been greatly facilitated by the availability of an already formalized core of knowledge contained in breast cancer protocols currently in use at the Fondation Bergonié. The experience reported by the Knowledge Systems Laboratory (7) on the development of the ONCOCIN system have been of great use in the design and implementation of the knowledge base. The development tool we used has proved to be efficient in the implementation and debugging phases, and apart from some limitations which can be corrected it provides most of the features that can allow cancer experts to develop useful, affordable and acceptable knowledge-based consultation systems.

The adjunction of temporal database management and statistical analysis capabilities to this kind of system can dramatically enhance the power and usefulness of such medical decision making tools, resulting in complete clinical research systems that could help in the management of clinical trials, and in the dissemination of state of the art knowledge acquired from clinical cancer research.

However, much remains to be done before one can anticipate the routine use of such systems in actual practice : formal evaluations of the systems have to be carried out in real world environment, improvements in the system design and features have to be done according to the results of evaluation studies. Finally, forensic and ethical issues should have to be addressed before these expert-systems could be widely used in clinical practice. An effective collaboration between cancer experts, laboratories involved in I.A. research and industry could greatly facilitate and expedite the achievement of these goals.

REFERENCES

1. BEGG CB, CARBONE PP, ELSON PJ, ZELEN M. : Participation of community hospitals in clinical trials : analysis of five years experience in the Eastern Cooperative Oncology Group. N Engl J Med 1982, 306, 1076-1080.

2. WIRTSCHAFTER DD, SCALISE M, HENKE C, GAMS RA : Information systems improve the quality of clinical research. Results of a randomized trial in a cooperative multi-institutional cancer group. Comput Biomed Res 1981, 14, 78-90.

3. LIBERATI A, ANDREANI A, COLOMBO F, CONFALIONERI C, FRANCHESCHI S, LA VECCHIA C, TOGNONI G : Quality of breast cancer care in Italian general hospitals. Lancet 1982, II, 258-260.

4. HICKAM DH, SHORTLIFFE EH, BISCHOFF MB, SCOTT AC, JACOBS CD : A study of the treatment advice of a computer-based cancer chemotherapy protocol advisor. Ann of Intern Med 1985, 103, 928-936.

5. KENT DL, SHORTLIFFE EH, CARLSON RW, BISCHOFF MB, JACOBS CD : Improvement in data collection through physician use of a computer-based chemotherapy treatment consultant. J Clin Oncol 1985, 3, 1409-1417.

6. RENAUD-SALIS JL, LAGARDE C, BONICHON F, AVRIL A, DURAND M : Developpement d'un système à base de connaissances pour l'aide à la décision thérapeutique dans les cancers du sein. Bull Cancer, in press.

7. SHORTLIFFE EH, CARLISLE SCOTT A, BISCHOFF B, BRUCE CAMPBELL A, VAN MELLE W, JACOBS CD : ONCOCIN : An expert system for oncology protocol management. In : Rule-based expert systems : the MYCIN experiments of the Stanford Heuristic Programming project, BUCHANAN BG, SHORTLIFFE EH : Reading, MA, Addison-Wesley, 1984, 653-668.

8. FOX J : Artificial Intelligence in primary care. Bull Cancer, in press.

9. RENAUD-SALIS JL, FOX J, LARGARCE C : Can cancer experts build expert systems ? Our experience in building an expert system intended to assist physicians in breast cancer management. 14th Intern Cancer Congress, Budapest, Août 1986 (+ poster).

Qualitative Reasoning

THE USE OF QSIM FOR QUALITATIVE SIMULATION OF PHYSIOLOGICAL SYSTEMS

Emma Nicolosi and Mark S. Leaning
Computer Unit, Royal Free Hospital School of Medicine
Rowland Hill Street, LONDON NW3 2PF

ABSTRACT

This paper describes an approach to the conceptual representation and qualitative simulation of compartmental physiological systems. Particular emphasis is given to the use of Kuipers' QSIM algorithm for qualitative simulation especially its reconstruction and some extensions. The work forms part of a project developing a new schema for representing dynamic biological systems on computer integrating conceptual, qualitative and quantitative approaches.
Keywords: Qualitative simulation, physiology, compartmental systems, knowledge-based modelling.

1 INTRODUCTION

Many aspects of the behaviour of physiological systems are described by human experts in terms of the dynamics of substances in discrete physiological spaces. Such a **compartmental** view is also the basis of numerous mathematical models [5]. This paper describes an approach to the conceptual representation and qualitative simulation of compartmental physiological systems. Emphasis is given to the use we have made of Kuipers' QSIM algorithm for qualitative simulation [7,8], especially its reconstruction in micro-PROLOG Professional running on IBM AT and some extensions which automate the initial phase of creating a qualitative description of a model (constraint network) from its physiological representation.

The work forms part of a project aimed at producing a new type of computer model for dynamic biological systems (the "MODEL" system, [10]). The QSIM algorithm has two important characteristics relevant to our project: it can work with incomplete knowledge of structure and underlying mechanisms; it produces the set of all possible qualitative time solutions of the model. In order to verify such features and to investigate in depth the advantages and disadvantages of QSIM, a Prolog reconstruction has been performed. Other goals are to improve QSIM and embed it within a new multilevel representation schema for biological systems, which includes conceptual, symbolic (qualitative), and numeric levels [9].

The intended applications of the new representation schema are to expert systems for physiological modelling, and for combining causal, qualitative and quantitative reasoning in medical diagnosis and treatment.

The paper presents in detail the structure of a qualitative model, and the rules for its automatic generation. We feel that this "internal" level of detail is more useful than general descriptions.

2 RELATED WORK

There are a number of approaches to the qualitative description and simulation of physical systems including Naive Physics, or Envisioning, developed by de Kleer and Brown [1,2], Qualitative Process Theory developed by Forbus [3,4], and Qualitative SIMulation (QSIM) developed by Kuipers [6,7,8]. All three approaches use conditional laws and a generate-and-test approach. Envisioning and Qualitative Process Theory are process-centered whereas QSIM is based on mathematical constraints abstracted from differential equations, which are not directly connected to a representation of the physical structure. An important feature of QSIM is its capability to discover new qualitative values of the system functions which correspond to a new overall system state ("landmark" values). Such a characteristic is unique to QSIM, and "guarantees to produce a qualitative behaviour corresponding to any solution to the original equation" [8]. For this reason, we selected QSIM as the algorithm for qualitative simulation with acceptable mathematical properties. However, to be useful to us QSIM had to be extended with a conceptual description language and an algorithm for generating the qualitative model as described below.

3 STEPS TO QUALITATIVE SIMULATION

The qualitative simulation described here is one of a number of features which are available in the MODEL system once a conceptual representation of the physiological system has been constructed [10]. Other features include the automatic writing of symbolic differential equations and numerical simulation. The steps leading to qualitative simulation with QSIM are as follows:
1. Constructing a conceptual representation, in terms of compartments, fluxes, losses and inputs;
2. Generating a qualitative model. Automatic selection of functions and constraints;

3. Deriving initial values, directions of change (i.q. values) and landmark
   values
4. Applying QSIM.

4 GENERATING A QUALITATIVE MODEL

Once the compartmental system has been defined conceptually as a set
of compartments, fluxes, losses, and inputs the QSIM network of constraints
is created. The network is based on the law of conservation of mass for
each compartment,

$$\text{netflow}_i = \text{inflow}_i - \text{outflow}_i \qquad (1)$$

(where inflow_i is the sum of all inputs and input fluxes into a compartment
i, outflow_i is the sum of all output fluxes from that compartment, and
netflow_i is the rate of mass gain.)

The computational complexity of QSIM grows very rapidly with the
number of functions and constraints in the qualitative model. The rules
given below have been designed to produce a minimal set of functions and
constraints to minimise the computational load.

4.1 Rules for Generating Functions

F1. for each compartment i in the system create a state function x_i
    representing the amount of substance;
F2. for each flux from compartment i to compartment j in the system create
    an f_{ji} function;
F3. for each loss from compartment i in the system create an f_{0i} function;
F4. for each linear flux create a k_{ji} function;
F5. for each compartment with more than one flux into it create an inflow_i
    function;
F6. for each compartment with more than one flux out of it create an
    outflow_i function;
F7. for each compartment with at least one flux into and one flux out of it
    create a netflow_i function;
F8. if the perturbation of the system is a constant infusion into
    compartment i create a u_i function.

4.2 Rules for Generating Constraints

C1. for each linear flux create a multiplication constraint
    $$\text{MULT}(k_{ji}, x_i, f_{ji})$$
C2. for each non linear flux create a monotonically increasing constraint
    $$M+(f_{ji}, x_i)$$
C3. for each compartment, if there is more than one input flux, then create
    $$\text{ADD}(f_{i1}, \ldots, f_{in}, u_i, \text{inflow}_i)$$

C4. for each compartment, if there is more than one output flux, then create

$$ADD(f_{0i}, \ldots, f_{ni}, outflow_i)$$

(Currently, the system is limited to not more than 2 fluxes in and out of each compartment.)

C5. <u>The Law of Conservation of Mass</u>

<u>If</u> there are many input and many output fluxes (i.e. $inflow_i$ and $outflow_i$ exist), then create

$$ADD(netflow_i, outflow_i, inflow_i)$$

<u>If</u> there are many inputs and a single output (f_{ji}), then create

$$ADD(netflow_i, f_{ji}, inflow_i)$$

<u>If</u> there is one input (in_i) and many outputs, then create

$$ADD(netflow_i, outflow_i, in_i)$$

(where $in_i = u_i$, constant infusion or f_{ij})

<u>If</u> there is a single input and a single output, then create

$$ADD(netflow_i, f_{ki}, in_i)$$

<u>If</u> there is no input in_i but many outputs, then create

$$MINUS(outflow_i, netflow_i)$$

C6. <u>Rate of Change</u>

<u>If</u> there is a single input and no output, then create

$$DERIV(x_i, in_i)$$

<u>If</u> there are many inputs and no outputs, then create

$$DERIV(x_i, inflow_i)$$

<u>If</u> there is one or more outputs, then create

$$DERIV(x_i, netflow_i)$$

5 DERIVING INITIAL VALUES, I.Q. VALUES AND LANDMARK VALUES

The system is considered in zero state prior to the perturbation with an input. Hence the <u>values</u> and <u>i.q. values</u> (incremental qualitative values or directions of change) of the functions x_i are derived as follows depending on the type of input:

	Impulse input		Constant Infusion	
	value	i.q. value	value	i.q. value
perturbed compartment	x_i^*	dec	0	inc
compartment directly linked to the perturbed one	0	inc	0	std
other compartments	0	std	0	std

The values and i.q. values of the other functions are subsequently derived from tables associated with the constraints in which the functions are present.

The set of landmark values of each function (f_i) depends on its initial value:
- initial value 0, $\longrightarrow$ {-infinity 0 +infinity}
- initial value $f_i^* > 0$, $\longrightarrow$ {-infinity 0 f_i^* +infinity}
- initial value $f_i^* < 0$, $\longrightarrow$ {-infinity f_i^* 0 +infinity}

6 APPLYING QSIM

Classical quantitative simulations are possible and useful when we want to simulate a well-defined numerical experiment on a model. When knowledge is incomplete (e.g. numerical parameter values or the exact mathematical description of a flux are not known) qualitative simulation can be used because it is symbolic (i.e. it does not need numbers) and it uses monotonic constraints to represent unknown relationships.

The QSIM algorithm is based on a generate-and-test approach. A set of possible transitionsA is attached to each function in the current state. The combination of transitions of different functions linked by a constraint is then tested for consistency with that constraint. Only the sets of transitions consistent with all constraints in the qualitative model will then define the successors of the current state.

The time dimension is organised as an alternate sequence of distinct time points t_i and intervals ($t_i\ t_{i+1}$). Therefore two types of transitions are possible: from t_i to ($t_i\ t_{i+1}$), and from ($t_i\ t_{i+1}$) to t_{i+1}. All solutions are stored in a tree of qualitative states, where a path in that tree represents one of the possible behaviours of the model. The simulation tree of the system is searched breadth-first in order to find all solutions without missing any alternative branches.

Once the QSIM simulation has been performed qualitative graphs of the behaviour of the functions are provided. For further details on QSIM consult Kuipers [7,8].

7 EXAMPLES

The examples presented show a typical constraint network and various qualitative simulations for a two compartment system shown in Fig.1. Such a model has wide physiological applicability, for instance in the description of drug kinetics (where 1 represents the drug in blood and 2 the drug in tissues).

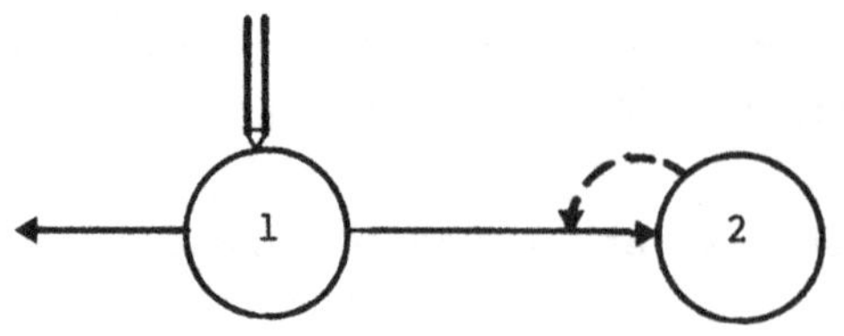

Figure 1. Two Compartment System

In Fig.1 it is assumed that the input in compartment 1 is an impulse (i.e an injection), the loss from compartment 1 is linear and the flux from compartment 1 to compartment 2 is nonlinear. However, the precise form of this nonlinear function is unknown.

The functions, values, i.q. values and constraints generated by our system are the following:

QSIM Network for the Compartmental System

functions	value	i.q. value	constraints
x1	+x1*	dec	DERIV (x1 netflow1)
x2	0	inc	DERIV (x2 f21)
k01	+k01*	con	M^+ (f21 x1)
f21	+f21*	dec	MINUS (outflow1 netflow1)
f01	+f01*	dec	ADD (f21 f01 outflow1)
outflow1	+out1*	dec	MULT (k01 x1 f01)
netflow1	-out1*	inc	

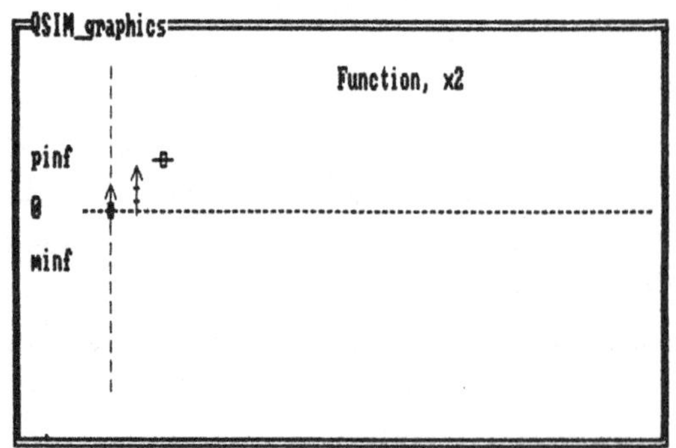

(where a square indicates a distinct point, two horizontal bars an interval, a single horizontal bar that the state is steady, and an arrow the direction of change)

Figure 2 Function x1 for 2 Compartment System

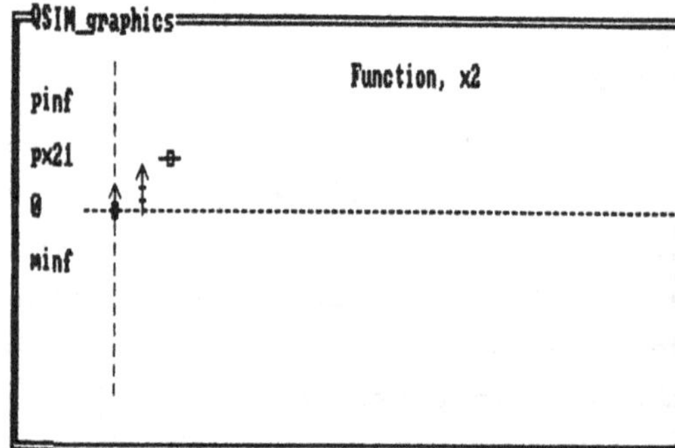

Two solutions are found by QSIM (i.e. the simulation tree has two branches). In the first (Fig.3) x2 reaches the steady state at +infinity (pinf), whereas in the second (Fig.4) x2 reaches steady state at a new landmark value +x21 (px21). x1 has the same behaviour in both the solutions (Fig.2). The qualitative graphs show complete agreement with the expected behaviour of such a system.

8 CONCLUSIONS

A critical and delicate phase of using the QSIM algorithm is in defining the initial state of the system. Kuipers [7,8] does not provide rules to follow during this phase. It is difficult to think in terms of first and second derivative as, for example, in the case of the i.q. value of the function $netflow_i = dx_i/dt$. Such a limitation has been solved in the MODEL system for compartmental systems [10] using the rules presented in Section 4.

QSIM is based on a generate and test mechanism: all transitions applicable to the current state are generated and then tested for constraint consistency. No mechanism is provided to limit the branching which results. The simulation tree grows rapidly and it becomes very large when we are dealing with systems of medium complexity. (In a 4 compartment model with unilateral fluxes, we have found a simulation tree with at least 16 million branches. The results are not shown.) The fact that the simulation tree is searched in breadth-first strategy does not improve the performance of QSIM, although it guarantees that all solutions are found.

Three types of solutions to the branching problem are being investigated: tree pruning; function reduction; constraint reduction. Tree pruning methods include: the use of mathematical theorems or other metaknowledge; heuristic search (e.g. weighting the branches of the tree); the use of numerical simulation results. Function reduction will be performed by using conceptual aggregation (e.g. combining compartments), whereas the use of more complex constraints (e.g. nonlinear saturation functions) will reduce the number of constraints necessary for the definition of the qualitative model.

Current work concerns the implementation of the above solutions together with the design of a new icon-based version of the MODEL system.

Acknowledgements

This work was in part supported by NATO Grant No RG 85/0207.

<u>REFERENCES</u>

1. de Kleer J and Brown JS.   Foundations of envisioning. In: <u>Proceedings</u> <u>National Conference on AI</u>. Pittsburgh 1982; 434-437.

2. de Kleer J and Brown JS. The origin, form and logic of qualitative physical laws. In: <u>Proceedings 8th IJCAI</u>. Karlsruhe, Germany 1983; $\underline{2}$: 1158-1169.

3. Forbus KD.   Measurement interpretation in qualitative process theory. In: <u>Proceedings 8th IJCAI</u>. Karlsruhe, Germany 1983; $\underline{1}$: 315-320.

4. Forbus KD.   Qualitative process theory. Artificial Intelligence, $\underline{24}$, 85-168, 1984.

5. Godfrey KR. Compartmental Models and their Application. London: Academic Press, 1983.

6. Kuipers B.   Commonsense reasoning about causality: deriving behavior from structure. Artificial Intelligence, $\underline{24}$, 169-203, 1984.

7. Kuipers B.   Qualitative simulation of mechanisms. Cambridge: Massachusetts Institute of Technology, Technical Report MIT/LCS/TM-274, 1985.

8. Kuipers B.   Qualitative simulation. Artificial Intelligence, $\underline{29}$, 289-338, 1986.

9. Leaning MS, Nicolosi E. Computer-representations of physiological systems: knowledge-based modelling. Technical Report 86/2, Medical School Computer Unit, Royal Free Hospital School of Medicine, London, 1986.

10.Leaning MS, Nicolosi E. MODEL: Software for knowledge-based modelling of compartmental systems. Accepted by Biomed. Meas. Inf. Contr., 1987.

Qualitative Description of Electrophysiologic Measurements:

towards automatic data interpretation.

W.J.Irler   (*)
R.Antolini   (+,*)
M.Kirchner   (*)
L.Stringa   (*)

(*)   I.R.S.T., I-38050 Povo - Trento , Italy
(+)   Department of Physics, University of Trento

Abstract:

The first steps towards an automatic interpretation of on-line measurements in intracardiac electrophysiologic tests consist in a transformation of the test data obtained by a beat-to-beat measurement device, in an adequate symbolic representation, an exhaustive description of the results and the discussion of inferable medical interpretation hypotheses. On the base of a simple test protocol, the cyclic interpretation sequence is pointed out with a survey over its subtasks. The qualitative description aspect is considered in detail. An example with knowledge representation issues of the used commercial shell and an output of the prototype version is presented.

Keywords: intelligent instrumentation, electrophysiology, interpretation

Introduction

One promising application of *Expert Systems* in medicine are interpretation support systems integrated in some medical measurement device. More or less operative approaches for so-called "smart instruments" treat blood chemistry measures (electrophoresis interpreter in [1], [2]), ECG, EEG or EMG tracings [3,4] or function test interpretation [5]. They work with assessed knowledge engineering concepts, are rule- or frame-based with an appropriate reasoning mechanism.

The expert system in this case operates mainly not in dialog form on user supplied answers, but has to reason pretty automatically about a set of measured data or an eventually continous data stream. To argue in some non-monotonic way about these values, a transformation in a symbolic representation is necessary on the base of which a forward- or backward-chaining of the rules built upon becomes possible.

Already the symbolic description terminology must pertain to a vocabulary used by the medical specialist. We will concentrate on the description aspects of the interpretation problem. Because of our intention to build the whole interpretation support from successive modules, we must elaborate these first aspects, the rest - perhaps - will be built on it.

On the way to our results, we will draw some attention to the general concepts involved in an automatic decision support system. The necessary distinction between consecutive and causal reasoning is attributed to the distinct behavioural requirements of expert systems: the reasoning and its "intelligent" explanation [6]. The need for a simple qualitative modeling is pronounced, more for explanation and justification than for the proper reasoning performance.

Some short preparatory remarks to the underlying medical knowledge sources - arrhythmology and electrophysiology - try to facilitate the understanding of our argumentation.

Protocols in Cardiac Electrophysiology

In the cardiac electrophysiologic test, the arrhythmologist tries to measure local electric potentials of the heart. To this pupose a catheter is introduced, the tip of which must be in contact with some heart region. A catheter may take the esophageal passage using the neighbourhood between the lower esophagus and the right atrium or enters via the vena cava to come to lie in the inner cardiac cavities. There may be more than one catheter. All catheters register local electric heart potentials, but are also used for pacing protocols [7]. A determined stimulation pattern gives the investigator indications for the etiology of an arrhythmia. Pacing is treatment, too, when the arrhythmic activity is interrupted. A drug administration during test tries to assess a probably effective chronic treament to prevent future arrhythmic episodes.

For demonstration we present the medical aspects only for a case with common atrial flutter studied with a transesophageal registration [8]. In atrial flutter the atrium beats with a higher frequency than the ventricle. A return to a regular rhythm and its maintainance is important for hemodynamic reasons.

Real-time Measurements

The time intervals between successive electric activations registered at the tip of the catheter is measured by a real-time measurement device [9]. Output of this device is a graphic with the beat-to-beat evolution of the intervals. Internally it is a sequence of data blocks memorised or transmitted for further elaboration in a file or a series of files. A record in a file contains a set of interval measurements, usually from atrial or ventricle leads (atrial-atrial cycles, atrium-His bundle conduction times, R-R intervals, etc.) and a corresponding set of validity flags. The the real-time aspect: one value set each 100 - 200 msec, is actually less important. In the future we will elaborate the data handling in a pipeline procedure to enable some parallelism, thus accelerating the system. Once the device is distributed for ß - test, a strict protocol observance is to be guaranteed, and a complete and comparable data set for each investigation case is mandatory.

The investigator needs furthermore a medical interpretation guide for the results with the patient still under examination, to decide a necessary protocol completion. A re-examination is often not thinkable within a short time and is very expensive.

Need for Interpretation

A medical diagnosis based on clinical measures requires a lot of extra-medical preparation of the data: statistics as well as graphics or images. Good laboratories print with each individual measured value its own valid normal ranges and underscore warnings in the case of non-normality. This is by no means an automatic interpretation, nor a medical diagnosis.The responsable doctor has to match the hints with his medical experience and then to come to a relevant judgement and diagnosis.

A pass forward to automatization is to introduce interpretations hidden in some complex pattern of different single values or temporally distributed measurements. In our case, we try to elicit the necessary information for a description of dynamically changing data, where a more complete medical interpretation with reasonable explanations may be the next pass.

Interpretation cycle

Like all medical examination the electrophysiologic test, too, has to perform a couple of evaluation - decision cycles going from measurement, interpretation, diagnostic hypothesis to needs for further measurements and so on. At the end, a more definitive conclusion like a therapeutically relevant diagnosis is reached or supposed.

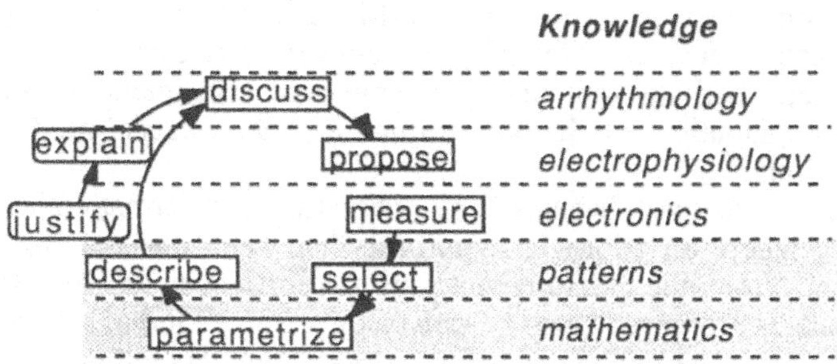

In figure 1., an idea is given, what singular steps are involved in the whole interpretation task. Furthermore these tasks are represented on an almost closed circle, to underline the analogue to a procedure where each item may be cyclically influenced by some predecessor. Each task - carrying the name of the desired action - is a context that has to be solved by the application of all attached rules before the next context may initiate. We expose here only the tasks that lie outside the straighter intent of our paper.

The task **"measure"** must hold all informations connected to the pure hardware/software elaboration of the signal , here intracardiac activation waves. It has to know the recognition technique of a chosen protocol. Its settings (thresholds, onset limits, peak intensity,..) may be subject to interactive adaptation caused by preceeding or initial values in the beginning phase.

The next element on the circle called **"select"** represents the inner structure of all used protocols. It guesses where in the whole file significant data is to be expected. These are single values like "the first spontaneous beat after pacing", or data blocks , thus representable by statistics. To obtain realistic mean values, however, some memory of the process is necessary; a new incoming point has to be attribuated to the right cluster depending on the preceeding ones: two always alternating distinct interval measurements (gallop rhythm) have a two-peaked distribution; a global mean must be substituted by the centers of each peak.

The element **"parametrize"** estimates parameters that describe the temporal distribution of the just selected representative points. Distinguable data clusters may collapse or split in more complicated patterns or simply become random. At this level, an adequate translation of values and parameters in a symbolic representation takes place. With **"describe"**, the proper symbolic elaboration begins with a local description of the previously obtained values (see next chapter).

The successive task in the cycle is **"discuss"** where, on the base of the exhaustive description, the medical (arrhythmological) interpretation of the test results is to be suggested. Here, the cycle contains some incongruity, caused by the necessary "how"-feature at this place, needed for an intelligent explanation. This latter must be to some extent causal. It is the medical background for the hypothesis that is requested, i.e. if in the discussion, an interpretation is given, like: "the drug decreases the atrial rate from 248 to 184 bpm", an explanation of this sentence has not simply to repeat the obtained mean values of one of the investigation phases nor the rules from this values to the symbolic representation. The investigator should hear something about the electrophysiological mechanisms, like: "the drug prolongs the atrial conduction" or "a reentry circuit is supposed", perhaps with an obstacle measure. We intend the sub-task **"explain"** to do this explanation job, having in mind a general causal model about arrhythmologic connections.

. The same happens for a successive "whence" , i.e. when further information is desired that should **"justify",** where an etiologic hypothesis had been derived from; the information hereto comes from the measurement procedure. An answer like: "the time between successive peaks is prolonged" may be satisfactory. Some kind of simulation is thinkable. We see, that these last two elements are general, not depending on specific values. They contain a qualitative model of electrophysiologic phenomena.

The acceptance of a clinical interpretation implies either a direct therapeutical decision and/or another test protocol. **"Propose"** is the element that should give suggestions in this direction. The cyclic continuation at this point is the decision of the investigator to switch to an other protocol.

Qualitative Description

Our interest in this presentation is to outline the description feature of the electrophysiologic test support. In fig.2. we see two short plots of atrial and ventricular heart rates of a patient with common atrial flutter [8]. A piece from the first minute (immediately before drug administration) and one during the tenth minute (during drug infusion) are confronted. For better visibility, the measured points are connected.

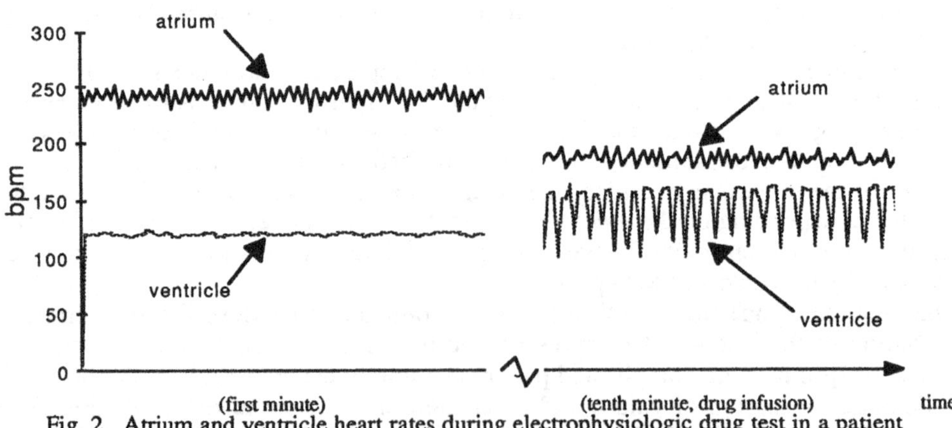

Fig. 2. Atrium and ventricle heart rates during electrophysiologic drug test in a patient with common atrial flutter (beat-to-beat plots for about 20 sec each).

The aspect on the upper left curve in fig.2. (atrial rate) resembles a pretty regular (up and down) sawblade form with a slight perturbation of lower frequency (about 2 sec long) superimposed. In this control phase the atrium shows one long, one short interval at twice the mean frequency of the relatively constant ventricle. Under drug effect - on the right side - the atrium becomes slower and more irregular. The ventricle shows a pattern of a few successive points one of which lies much lower, the others near the atrial rate.Statements like those are the kind of description needed for a medical interpretation. Remember that they only describe the aspect of the graphic using merely the names of its content. If we are able to deliver a similar satisfactory description, we have reasonable hopings to be able to build an automatic interpretation tool, too.

In fig.1. we had postulated "patterns" as the knowledge source for "describe". That means: apart from knowing the denomination of the signals, only general considerations about data distributions and phenomenological aspects play a role.

Knowledge implementation

In addition to our first presentation of the interpretation support system [10], we intended to experience the description problem with a commercial expert system shell that is based on work in medical decision making [11]. On the one hand for representational generality and on the other to discover needs that are not covered.

We don't know a recent knowledge engineering tool that could be regarded as apt to manage continous data streams, even without taking into account the real-time problem. Temporal reasoning is still an unresolved research theme in knowledge engineering [12,13]. In the knowledge elicitation phase and the first experiments with an operational representation, a general shell seems, however, pretty well suited.

Our experimental knowledge base contains about 50 rules with prevalent domain-specific informations. Some others have more control functions and are not interesting here. All rules make excessive use of a special kind of attribute generalization: the so-called attribute classes [14]. These permit a definition of attributes for a whole class the instantiations of which must not be pre-fixed. Class hierarchies are not possible. With the classes, the majority of rules are valid for all instances. This feature allows to construct a relatively compact knowledge base with few redundancies. A small segment of rules that determine the overall rough form of a series of cycle measurements is:

```
Form zigzag:
s:steps
if        s>Course      = up_down : down_up
then      s>Form        = two_valued <0.7>.
endif.
```
"if the steps course follow an up-down or down-up sequence, then their form will be two-valued."

```
Cycle form:
c:cycle, s1:steps, s2:steps
if        c>Name       = actual:Cycle>Name and
              s1       ≠ s2     and
          s1>Form      = s2>Form
then      c>Form       = s1>Form.
endif.
```
"look over all cycles and steps; bind c to the cycle actually under consideration; if two non-identical steps present the same form, accumulate its evidence to that of cycle c.

The second rule would have been written in conventional programming style (a double loop), without the class definition feature. Nevertheless, its first premise is a control structure; the reason: for this problematic, a further, not provided hierarchical level would be needed.

Results

Up to now, we completed the rules for a few protocols. Our description of the drug test data requires a micro elaboration time which is relatively long (over 1 second for one data record - 3 to 4 times the generation time). But there is good hope to remain in the total test period, because the tests take normally more than half an hour, and only small, selected data segments have to be considered in the first guess. In the case of problems, however, the need for other data blocks could create a relevant delay. This is not our actual center of interest.

A typical output of a test is:

> In **T1** [sec **3** to **25**], the form of **A-A** is **two_valued** **<0.98>**.
> The upper mean is **254** bpm (min: **250**, max: **264**);
> the lower mean is **248** bpm (min: **238**, max: **252**)
> A **slight** **<0.8>** perturbation (cycle length **2.4** sec) appears in the atrial pattern
>
> The evolution of the A-A rate from t=**3** to t= **568** sec is **decreasing** **<0.6>** from **248** to **184** bpm.

Calculated and infered values are in bold type. Angular brackets enclose the corresponding evidence measures.

Discussion

Each physician has learned to produce a detailed case description for all examinations he does. The proper medical hypotheses about diagnosis have to appear at the end and to figure as thus, i.e. stated with a more or less explicit question mark. The same "good scientific practice" requires for clinical studies at least an exploratory data analysis [15], a data description in statistical terms, frequently the main part of a statistical elaboration.

We, too, tried to demonstrate, that it is a solid qualitative description that does the main job in an intelligent instrumentation. For our electrophysiologic test support system we aspire by this way some not negligible intermediate results, namely:
- *compress the data,*
- *motivate the collaboration of the investigator,*
- *facilitate the participation in the reasoning,*
- *direct, but not anticipate an interpretation,*
- *create the symbolic fundament for automatic interpretation.*

An other argument seems just so important: perhaps <u>the physician does not want a perfect interpretation</u> delivered by the medical instrumentation. Such an objection has to be taken into account and considered in the right way. Our main goals wanted only to
- *deliver an orientation in a novel instrumentation,*
- *assure complete protocol observance,*
- *memorize the reactions to the delivered propositions.*

These objectives are satisfied with our first passes. In view of the fact, that the underlying domain, electrophysiology, does not yet possess an accepted causal theory - an essential for a knowledge representation -, we think, a somewhat consciously incomplete interpretation, a "mere" description plus some hints may be of more interest than a - perhaps artificial - definit solution. With these arguments, we feel nearer to the signification of interpretation *support* system and to the consideration of medical expert systems as serious partners of the responsable physician for a "curbside consultation" between "friendly colleagues".

Acknowledgements

We wish to express our thanks to Prof.F.Furlanello and Dr.M.Disertori from the Arrhythmologic Center of the St.Chiara Hospital, Trento, Italy for the kindly permission to use the clinical data.

References

[1] Weiss S. and Kulikowski C. (1984), A Practical Guide to Designing Expert Systems. Rowman & Allanheld Publishers: Totowa NJ.

[2] Besset D. (1986), BABAR: an expert system for blood analysis. *First Swiss Symposium on Medical Informatics*, Telmed ed: Ruschlikon.

[3] Shibahara,T. (1985), On using causal knowledge to recognize vital signals: knowledge-based interpretation of arrhythmias. *Proc.Int'l Joint Conf.Artificial Intelligence*, 1985.

[4] Baas L. and Bourne, J.R. (1984), A rule-based microcomputer system for electroencephalogram evaluation. IEEE *Trans.on Biom.Engin.*,vol.BME-31, 10, Oct.1984.

[5] Aikins J.S., Kunz,J.C. and Shortliffe,E.H..(1983), PUFF: an expert system for interpretation of pulmonary function data.*Computers and Biomedical Research*, vol.16, 199-208.

[6] Clancey W.J. (1983), The epistemology of a rule-based expert system - a framework for explanation. *Artificial Intelligence*, 20, 215-251.

[7] Josephson M.E. and Seides S.F. (1979), Clinical Electrophysiology. Techniques and Interpretations. Lea & Febiger: Philadelphia.

[8] Disertori M.,Vergara G.,Inama G.,Guarnerio M.,Antolini R., Furlanello F. (1985), Electrophysiologic effects of flecainide in atrial flutter. Acute test evaluation by on-line data analysis. *New Trends Arrhyt.*,1(3),255-259.

[9] Antolini R., Kirchner M., Mongera A., Disertori M.,and Furlanello F. (1984), Real-time beat-to-beat measurement of conduction intervals during cardiac electrophysiologic studies. *Clin.Phys.Physiol.Meas.*, 5(3), 171-183.

[10] Irler W.J., Antolini R., Kirchner M. and Avancini G.P. (1986), On the application of expert system concepts to on-line interpretation of intracardiac electrophysiologic testing. *Measurement in Clinical Medicine*, IMEKO : London , 153-158.

[11] Reggia J.A. (1982), Computer-assisted medical decision making. *Applications of Computers in Medicine*, IEEE.

[12] Allen J.F. (1983), Maintaining knowledge about temporal intervals. *Comm.* ACM, vol.26, 11,822-843.

[13] Tsotsos J.K. (1981), Temporal event recognition: an application to left ventricular performance evaluation. *Proc.Int'l Joint Conf.Artificial Intelligence*, 1981.

[14] KES manual , ver.2.2 (1986), Softw.Arch. & Engin. Inc.: Arlington VI.

[15] Tukey J.W. (1977), Exploratory Data Analysis. Addison-Wesley: Reading, Mass..

A QUALITATIVE SPATIAL REPRESENTATION FOR CARDIAC ELECTROPHYSIOLOGY

Nick Gotts

Artificial Intelligence in Medicine Group
School of Engineering and Applied Sciences
University of Sussex
Brighton, BN1 9QT

ABSTRACT

A qualitative representation for the distribution of polarized and depolarized areas of conductive tissue in the heart is presented. This is based on the idea of spatially continuous "lobes" of tissue made up of cells which are all in a similar electrical state. The range of possible lobe-configurations is investigated. Changes in lobe-configuration are discussed in relation to qualitative states of lobe-boundaries. Pointers to possible uses of the representation are given.

Keywords: qualitative representation, spatial reasoning.

1. INTRODUCTION

This paper reports an ongoing attempt to extend AI work on qualitative modeling [1,2,3,4] to the domain of spatial reasoning. Qualitative spatial reasoning has so far made little progress. The question addressed here is whether useful qualitative representations can be developed for the spatial aspects of changing electrical states within the heart. This is part of an attempt to represent the mechanisms of cardiac arrhythmias as understood by cardiologists.

In representations for qualitative modelling, a complete qualitative state generally includes two distinguishable types of information, which can be called "snapshot" information (or "qualitative *state*" information in a narrow sense), which specifies which of a finite set of states or configurations the system currently occupies; and "dynamic" information, concerning the types of continuous change currently occurring, and tending to take the system into another state. This paper concentrates on "snapshot" information, although sections 5 and 6 concern "dynamic" information.

2. ELECTRICAL BEHAVIOUR OF THE HEART

When the heart is relaxed, the cells of the myocardium (heart muscle), and of the specialized conducting system running through it, are polarized: there is a voltage across the cell membrane. Depolarization causes muscular contraction.

Depolarization is a "domino" process: depolarized cells influence their still-polarized neighbours to depolarize in turn. A depolarization wave begins when one or more cells depolarizes spontaneously. Many heart cells will do this given sufficient time, but depolarization normally begins in the sinus node. This is a structure in the right atrium containing cells which are particularly quick to depolarize spontaneously. The wave spreads thence through the conductive parts of the heart. Normally, its progress will be continuous; but it may cease and then resume its advance, if it encounters cells not ready to depolarize.

Repolarization follows depolarization, preparing the heart for the next depolarization wave when it arrives. Repolarization is not a "domino" effect like depolarization. The time at which a depolarized cell repolarizes is decided by when it was depolarized, its intrinsic rate of recovery, and conditions in the local extracellular environment. Nevertheless, there generally *appears* to be a coordinated repolarization "front", with nearby cells repolarizing close together in time. This is a result of the distribution of cells' intrinsic rates of recovery in the normal heart. It is assumed here that repolarization is a spatially orderly process, but that a repolarization "front" may, like a depolarization wave, become static for a short but significant period and then resume its movement.

3. LOBES AND LOBE-BOUNDARIES

The approach explored here starts with the idea of dividing a conductive piece of heart tissue into parts ("lobes"), in such a way that each lobe is spatially continuous, and consists of cells in a single electrical state, differing from the state of cells in any neighbouring lobe. Lobes are of only two types: polarized ("P-lobes"), and depolarized ("D-lobes"). A D-lobe's neighbours must therefore be P-lobes, and vice versa. It may be necessary later to introduce other types of lobe, in particular a type to represent areas of chaotic electrical activation. However, the restriction to two lobe-types considerably reduces the range of possible lobe-configurations which can arise.

The approach described could be applied either to the electrically-conducting parts of the heart as a whole, or to some smaller anatomical region such as the sinus node. The area to be modelled will be referred to here simply as the "region". Topologically different regions allow different ranges of lobe-configurations. The simplest type of region is a solid piece of tissue, without holes through it (such as a quoit or ring-doughnut has), or interior voids (as within a tennis ball). This type of region is referred to here as a "solid-ball region", or simply a "ball region". The conducting region of the heart is topologically moderately complex, so analysis needs to extend to other types of region before a model of the complete heart could be built using the representations developed below.

The way a region is divided into lobes (by "lobe-boundaries" between adjacent pairs), could be described at various levels of detail. At one extreme, it could simply be specified that there are so many P-lobes, and so many D-lobes, within the region. At the other extreme, the topological properties and relationships of each lobe and lobe-boundary could be fully specified. Here, the focus is on a relatively coarse level of description: specifying how many D-lobes and P-lobes there are, and how many lobe-boundaries (if any) each pair of lobes shares. However, in order to analyse the range of lobe-configurations which can arise in any given type of region, it has proved necessary to consider the possible topologies of lobe-boundaries as well as their number.

A pair of lobes will not be allowed to touch only at a zero-dimensional point or along a one-dimensional line, as would be possible if they were thought of as arbitrary three-dimensional forms able to take up arbitrary spatial relationships. To be precise, it is assumed that any lobe-boundary between two lobes consists of a two-sided two-dimensional surface. A lobe-boundary may be, in topological terms, either a closed (edgeless) surface - a sphere or an n-hole torus; or an open surface, with one or more edges, such as a disc with one or more holes in it. Each edge of an open-surface lobe-boundary will be a closed curve. The insistence that a lobe-boundary should be a two-dimensional surface simplifies analysis of the range of possible configurations considerably - and should also simplify the computation of possible configuration-changes. It is an idealisation of the real situation in the heart, where the geometrical and functional relationships between nearby cells are complex; but no more so than the assumption that the conductive tissue is infinitely divisible, which would be needed if lobes were permitted to meet along sharp, one-dimensional lines.

If a lobe-boundary is a closed surface, one lobe will be *inside* the other, and this will be the only lobe-boundary between the two. An open-surface lobe-boundary may be one of several shared between the same pair of lobes, as in the configuration shown in cross-section in Figure 1. The two lobe-boundaries shown here are both discs, seen from above. (In this figure as in some others in the paper, two dimensions are used to represent three. All regions and lobes are shown as rectilinear: this is purely for convenience. Shaded areas represent holes through the region or voids within it. With the exception of Figure 5, assume that all the lobes and holes shown extend through the region from top to bottom, retaining the same cross-section.)

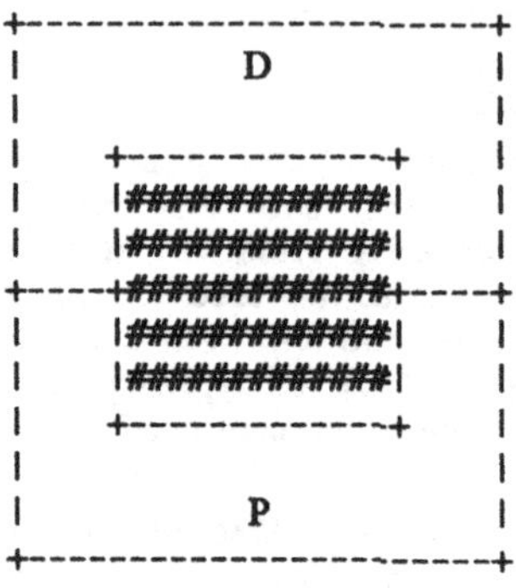

Figure 1

4. LOBE-CONFIGURATIONS

If no limit is placed on the number of lobes a region can contain, there will be an infinite number of possible lobe-configurations. In practice, the number of lobes present at any one time even in the whole heart is most unlikely to exceed 10. It will therefore be reasonable to place fairly low limits on the number of lobes in the region, to produce a useable model. Exactly what the limit should be can be decided independently for each model developed using the lobe-configuration representation.

With only two possible lobe-types, no trio of lobes can be mutually adjacent. This is a special case of a more general constraint, best explained by translating the problem into graph-theoretical terms. Representing each lobe by a graph-vertex, and a shared lobe-boundary by the existence of a link between two vertices, the two types of lobe can be represented by giving each graph-vertex one of two colours, and insisting that only vertices of different colours can be joined by a link. This means that not only can there be no trio of mutually adjacent lobes, but there can be no "ring" consisting of an odd number of lobes, each adjacent to its two neighbours around the ring.

In a solid-ball-region, as will be informally proved below, the graph of lobe-adjacency must be a tree - that is, free of loops altogether. Nor can there be two links between a pair of vertices, which would represent a pair of lobes with two lobe-boundaries in common. This can be regarded as a special type of loop in the graph.

Assume that there *is* a way of dividing a ball-region into lobes in such a way that the graph of lobe-adjacencies contains a loop. This implies that there is a sequence of lobe-boundaries, one between each successive pair of lobes around the loop, such that a closed path can be drawn which crosses each lobe-boundary in turn, without crossing any of them twice.

Consider the properties which these lobe-boundaries must have. A *closed* lobe-boundary cannot be part of such a lobe-loop. A closed lobe-boundary exists only when one lobe completely *surrounds* another - and it is always the only lobe-boundary between the two, constituting the outer surface of the surrounded lobe, and an inner surface of the other. Suppose the outer lobe is a D-lobe ("D1") and the inner a P-lobe ("P1"). If D1, P1, and their lobe-boundary were part of a lobe-loop, there would be a closed path beginning in D1, crossing into P1, and returning to D1 either via a second lobe-boundary between the two, or via other lobes. We know there cannot be another lobe-boundary between D1 and P1, so the second case is the only possibility. However, any lobe-boundaries which P1 has with other lobes must be *internal* to P1, since its entire external surface is the closed lobe-boundary with D1. If the path from D1 crosses one of these internal boundaries of P1, entering a second D-lobe ("D2"), there is no way it can return to its starting point in D1 without again passing through the D2-P1 and P1-D1 boundaries. So, no closed lobe-boundary can be part of a lobe-loop.

If a given lobe is to be part of a lobe-loop, it must therefore have at least two *open* lobe-boundaries. Given that there are only two types of lobe, each of these lobe-boundaries must have its edge or edges lying in the surface of the region itself. For imagine that any part of one of these edges were in the *interior* of the region. To one side of the edge would be the lobe-boundary, and to either side of that, a P-lobe and a D-lobe; but what would lie *beyond* the edge of the lobe-boundary? (See Figure 2a.) If there were a third type of lobe, the edge could be shared between three lobe-boundaries, as illustrated in cross-section in Figure 2b (where the edge is represented by the meeting of three lines in the middle of the region). With only two lobe-types, the edge must, as stated, lie in the region's surface.

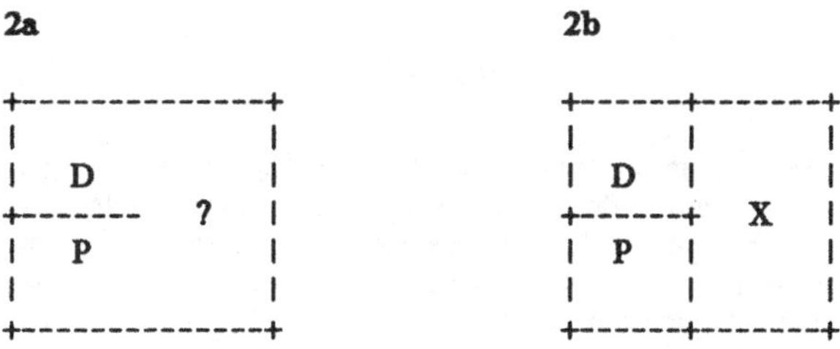

Figure 2

In a ball-region, however, an open surface with its edge(s) lying in the region-surface necessarily divides the region into two separate pieces, such that any path through the region from a point in one piece to a point in the other, must pass through the surface. A path running from a point in the lobe (say, D1) on one side of such a lobe-boundary, across the boundary into its neighbour (say, P1), can only be closed by recrossing the same lobe-boundary in the opposite direction.

Notice that this is not so for a doughnut-shaped region: as can be seen in Figure 1, a lobe-boundary can be created which cuts right through one side of a doughnut, in such a way that there are still paths through the region which connect points on either side of the lobe-boundary, without crossing it. Returning to the ball-region we are concerned with, the fact that an open lobe-boundary cleaves the region in two rules out the existence of lobe-loops - and also ensures that any pair of lobes in such a region can have only a single lobe-boundary, whether closed or open.

The adjacency-graph for lobes within a solid-ball region must therefore be a single tree, with two different types (or "colours") of vertex, which cannot be adjacent to each other. The adjacency trees of possible configurations with up to five lobes are illustrated in Figure 3.

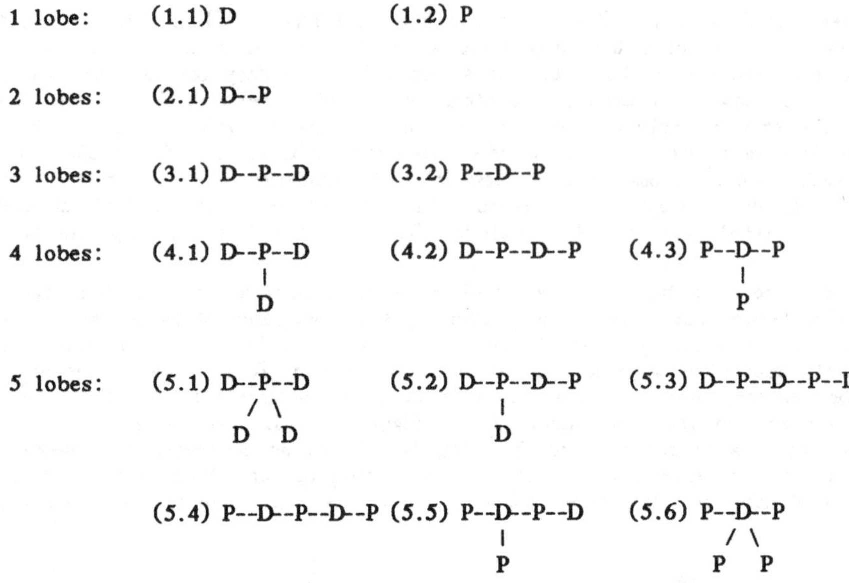

1 lobe: (1.1) D   (1.2) P

2 lobes: (2.1) D--P

3 lobes: (3.1) D--P--D   (3.2) P--D--P

4 lobes: (4.1) D--P--D   (4.2) D--P--D--P   (4.3) P--D--P

5 lobes: (5.1) D--P--D   (5.2) D--P--D--P   (5.3) D--P--D--P--D

                 (5.4) P--D--P--D--P (5.5) P--D--P--D   (5.6) P--D--P

Figure 3

In a doughnut-shaped region, as already indicated by Figure 1, the graph of lobe-adjacency can include a loop. A two-hole "doughnut" allows two non-overlapping lobe-loops, as indicated in cross-section in Figure 4. Clearly (since this pattern of lobes could be extended indefinitely to right or left), an n-hole doughnut will allow n disjoint lobe-loops, but we have not shown that this is a maximum.

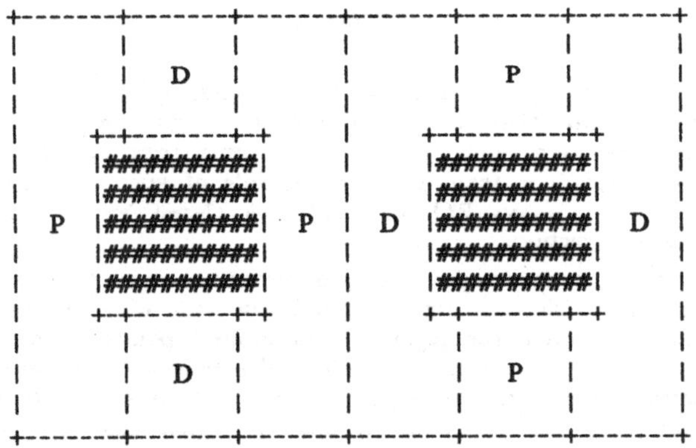

Figure 4

Voids within the region may also permit the existence of lobe-loops. A ball-shaped void does not appear to do so, as it does not permit the existence of any additional non-region-dividing lobe-boundaries. On the other hand, a doughnut-shaped void does do so. Consider a region consisting of a ball with a doughnut-shaped void within it (like an apple in which a maggot has eaten out a circular track around the core). The uneaten part of the apple can be divided into four lobes, which form a lobe-loop. Figure 5a shows a cross-section in the plane of the void: D1, D2, and the void can all be visualised as having the same thickness, together making up a horizontal "slice" through the cuboid "apple". P1 can then fit above this slice, and P2 below it, creating the lobe-loop D1, P1, D2, P2. Figure 5b shows a vertical cross-section.

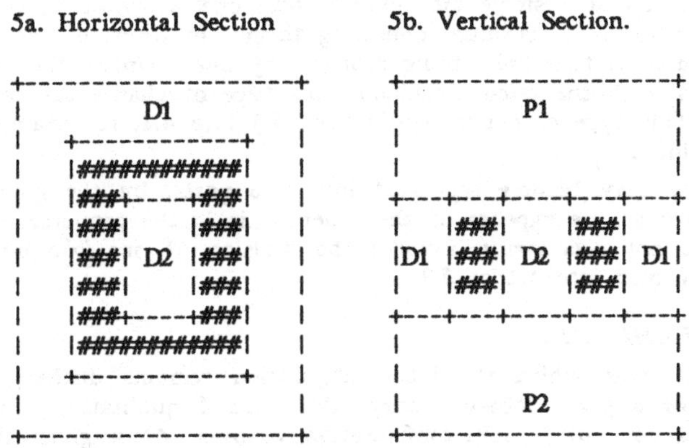

Figure 5

It is conjectured that the total number of disjoint lobe-loops a region permits can be calculated from the connectivity of its surfaces: a spherical surface contributing nothing, while each external or internal surface in the form of an n-hole torus permits an additional n disjoint lobe-loops. This has not been proved.

5. REPRESENTING CHANGES OF LOBE-CONFIGURATION

The first point to make here is that a set of *minimal* lobe-configuration changes should be sought, excluding any type of change which can be brought about by a sequence of two or more smaller qualitative changes. Non-minimal changes can then be composed of sequences of minimal ones. (The "sequence" of changes need not be assumed to be strictly ordered temporally: "successive" changes in such a sequence could be permitted to occur simultaneously, but not in reverse order.) The second point to bear in mind is the underlying conceptual basis of the changes which it is intended to model: changes in the lobe-configuration being conceived of as resulting from changes in the electrical states of cells within the lobes. It simplifies the task of finding a set of minimal qualitative changes to assume that such a change involves a set of cells, all of which are within a single pre-change lobe; and that all these cells are again in a single lobe (of the opposite type) at the end of the change.

There are four types of event which could bring about such a minimal change in the lobe-configuration within a region:

(1) A new lobe may appear somewhere within a lobe of the opposite type. The new lobe would therefore have a single neighbour. In the graph representation, this change would be represented by adding a vertex for the new lobe, and a link between that vertex and the one representing the old lobe within which it appeared. For example, configuration 4.1 in Figure 3 could become 5.1 by the appearance of a new D-lobe, or 5.2 by adding a P-lobe.

(2) A lobe *with a single lobe-boundary* may disappear. In graphical terms, this would involve the removal of a vertex, and of the link between that vertex and its single neighbour - turning 5.1 into 4.1, or 5.2 into either 4.1 or 4.2. The disappearance of a lobe with more than one lobe-boundary can be broken down into two or more smaller changes,

producing one or more intermediate qualitative states. Considering a lobe with two lobe-boundaries, the first step would be the replacement of part of that lobe by a connecting isthmus between the two - see (3) below. This would leave the diminished lobe with a single lobe-boundary. The second step would be the disappearance of the lobe.

(3) Two lobe-boundaries of a single lobe may become one: a switch in state among some of the its cells can form an isthmus connecting them. In terms of the graphical representation, this results in two links being replaced by one. Unless the two lobe-boundaries were boundaries with the same neighbour, this type of change also results in two *lobes* becoming one. This type of change could turn 5.3 into 4.1, for example, by amalgamating the two P-lobes.

(4) A lobe-boundary may become separated into two parts, by the growth of one of the lobes it separates at the expense of the other. This is the converse of (3). This change will often, but not necessarily, involve the division of one *lobe* into two. Thus 4.1 could be transformed into 5.2 or 5.3.

6. DIRECTIONS OF CHANGE

During the intervals over which the lobe-configuration remains unchanged, lobe-boundaries will be moving. For a given lobe-boundary, there are 4 qualitatively distinct states with different implications for future lobe-configuration changes. At a given time, depolarization may or may not be taking place across some part of the lobe-boundary. Similarly, repolarization may or may not be occurring. Since neither need be happening at a given boundary, and conversely both can go on at once across different portions of it (see below), there are 4 possibilities. A lobe-boundary change-of-state occurs when either depolarization or repolarization starts or stops across a lobe-boundary.

Depolarization and repolarization may occur simultaneously across a single lobe-boundary as a result of a local hiatus in repolarization. Imagine a region which is wholly depolarized. Repolarization begins, and at first spreads smoothly, so that a new P-lobe is expanding at the expense of the D-lobe all along their boundary. At some point, part of the repolarization front reaches an area which is not yet ready to repolarize. Here, the D-P boundary will stop moving into the D-lobe. It is then possible for cells on the D-side to trigger their repolarized neighbours into a new depolarization, setting part of the boundary moving in the opposite direction. This may take place while other parts of the boundary are still moving in the original direction.

The qualitative state of a lobe-boundary puts limits on possible changes in lobe-configuration involving the lobes on either side. Possible appearances of new lobes are not affected, as these do not appear at existing lobe-boundaries. However, a lobe which is advancing across any part of one of its boundaries cannot disappear. (Such a lobe could be undergoing a net shrinkage, but a lobe is growing across part of a boundary only if some cells just across the boundary are switching to its state - and these cells will remain in their new state for a significant period of time. Thus it is reasonable to insist that all growth of a lobe should stop before it can vanish.) If a lobe is to split into two, it must be retreating across at least part of its boundary with the lobe that splits it; and conversely, if two lobes are to join, at least one of them must be advancing across some part of its boundary with the third lobe separating the two.

What makes these connections between lobe-boundary-states and changes in lobe-configuration potentially useful is that, while it is *possible* for a lobe-boundary to alter its state, this is not common. Generally, a lobe-boundary comes into existence when a new D-lobe or P-lobe appears, and advances smoothly until the lobe it is expanding into has disappeared.

7. MAKING USE OF THE LOBE REPRESENTATION

To be used in working qualitative models of the heart, the single-region lobe-based qualitative description scheme outlined needs to be augmented, to allow multi-regional models to be used, and to include the qualitative description of temporal relationships (for which a version of Allen's representation [2] is to be used).

In discussing heart rhythms, cardiologists divide up the conduction system of the heart anatomically, and frequently concentrate attention on events within a single anatomical region. In a multi-regional model, individual lobes could extend into two or more regions at once. A normal heartbeat could then be described as a state-sequence, beginning with a state in which one P-lobe covers all regions. The next state would be one in which a D-lobe shared the sinus node with the P-lobe, with depolarization going on at the lobe-boundary. Later states would see the D-lobe moving into successive regions. At some stage, a new P-lobe would appear within the sinus node and begin to spread in its turn.

ACKNOWLEDGEMENTS

Thanks to Jim Hunter for helpful comments. This work was funded by SERC under Alvey grant GR/D/1827.1. This paper includes material from a paper entitled:

"The Qualitative Description of Spatial Aspects of Cardiac Electrophysiology"

which is to be published in "Biomedical Measurement Informatics and Control", the journal of the British Medical Informatics Society.

References

1.   B.J. Kuipers, "Qualitative Simulation of Mechanisms," MIT/LCS/TM-274, 1985.

2.   J.F. Allen and H.A. Kautz, "A Model of Naive Temporal Reasoning," in *Formal Theories of the Commonsense World*, ed. J.R. Hobbs and P.C. Moore, pp. 251-268, Ablex, Norwood, New Jersey, 1985.

3.   K.D. Forbus, "Qualitative Process Theory," *Artificial Intelligence*, vol. 24, pp. 85-168, 1984.

4.   I. Mozetic, I. Bratko, and N. Lavrac, "The Derivation of Medical Knowledge from a Qualitive Model of the Heart," Report, The J. Stefan Institute, Ljubljana, Yugoslavia, 1984.

Knowledge Acquisition and Representation

Knowledge acquisition in
expert system assisted diagnosis :
a machine learning approach

M. Funk[1], R. D. Appel[1], Ch. Roch[1],
D. Hochstrasser[2], Ch. Pellegrini[1], A.F. Müller[2]

[1] *Centre Universitaire d'Informatique,*
12, rue du Lac, Genève, Suisse
tel : (022) 87 65 80
[2] *Clinique Médicale, Hôpital Cantonal Universitaire,*
Genève, Suisse

Keywords
knowledge acquisition, medical expert systems, machine learning.

Abstract

Two-dimensional gel electrophoresis is a biochemical technique for protein separation, producing complex images containing thousands of spots. Their expert system assisted interpretation would be of great use in a clinical environment. Nevertheless, there are currently no experts capable of analysing these images. A machine learning system is presented that allows the expert system to acquire new knowledge and to formulate rules to diagnose new diseases from such a picture. A heuristic clustering method is used to group the images into separate classes and to determine the typical spots for each disease.

1 Introduction

MELANIE (Medical ELelectrophoresis ANalysis Interactive Expert system) is an interactive computer system to assist in the analysis and interpretation of two-dimensional gel electrophoresis images, that has been developed with the goal of helping in the setting up of a diagnosis [Fun87,App87].

Two-dimensional electrophoresis is a biochemical method consisting of separating proteins on a first gel according to their isoelectric point, and then on a second rectangular gel on which the first one is placed, according to their molecular weight. We can, in this way, obtain a two-dimensional map of the analysed proteins. It is possible to separate more than 3000 different proteins in a tissular or cellular sample of a few milligrams (fig. 1).

An expert system capable of interpreting such images, i.e. possessing the necessary knowledge for detecting abnormal spots, would be very valuable during the diagnosis.

Nevertheless, there are currently no medical experts capable of analysing these images in an effective way. Building an expert system will thus be hindered by the lack of domain-specific knowledge.

In order for the expert system to be more than an academic toy, i.e. to be usable in a clinical environment, it has to possess more knowledge than the physician. This is only possible if automatic knowledge acquisition facilities are incorporated into the system.

A machine learning system has therefore been developed in order to give the expert system the ability to formulate new rules from a limited number of images. The aim is to distinguish between several two-dimensional gel electrophoresis images, without knowing if or how many classes have to be formed for setting up a more precise diagnosis.

2   A heuristic clustering algorithm for two-dimensional gel electrophoresis images

The task of the learning system may be described as follows :

> Given that $V = \{g_1, g_2, ..., g_s\}$, a set of s images of two-dimensional gel electrophoresis from patients suffering from a disease O,
>
> find $P = \{P_1, P_2, ..., P_k\}$ a partition of V, and a function $Q(g)$, so that

$$Q(g_i) = j, \quad \forall g_i \in P_j,$$

> indicating that gel g_i belongs to class j.

This system should therefore initially classify a set of two-dimensional gel electrophoresis images into sub-classes, then create rules to determine which of the sub-classes a new gel g $(g \notin V)$ belongs to.

The motivation for developing this function of the learning system comes from the assumption of existing sub-categories not yet detected among certain types of diseases. One will give, as example, hepatitis non-A non-B which is the general name of a disease for which the number of viruses or toxic agents causing it has been unknown up to the present.

It is necessary, at this stage, to mention what entities characterize the problem of the classification of two-dimensional gel electrophoresis images. The following information is available :

- **n**, the number of two-dimensional gel electrophoresis images to classify;

- **k**, the number of classes to form;

- **s**, the number of variables defined for the objects to classify, i.e. the number of spots (proteins) contained on the images of two-dimensional gel electrophoresis;

- **o**, the number of distinct values possible for the intensity of each spot.

In addition, it is interesting to consider :

- **t** (t $\leq$ s), the number of proteins characteristic of a class, i.e the number of spots taking distinct values for o within distinct classes.

Moreover, several particularities of two-dimensional gel electrophoresis images should be mentioned :

- **the large number of variables.** As a matter of fact, each gel contains between 300 (serum, plasma) and 3000 (cells) different proteins.

- **the difficulty of determining the values taken by the variables.** Every protein is determined by the position of the corresponding spot on the gel, either according to its isoelectric point and its molecular weight, or according to the pixel coordinates of the image in X and Y. However, the imprecision of the biochemical technique complicates the process of spot identification.

- **the combinatorial explosion.** Let, for instance, the values for the symbols introduced above be: $n = 10, k = 3, s = 3000, o = 10, t = 1$. In other words, we have 10 gels, each containing 3000 spots, whose intensity varies, for instance, from 0 to 9. We want to form 3 classes each having one single typical spot. This means that we are trying to obtain the following information :

1. Which class each gel belongs to,

2. Which spot is characteristic of each of the three classes,

3. Which values of intensity this spot may take for a gel to belong to the corresponding class.

A quick calculation shows that the number of possibilities to form three classes by determining the three above-described pieces of information is in the region of 10^{16}.

The clustering algorithm that has been implemented is an adaptation of the one proposed by Michalski and Stepp [MS83]. The general structure is as following :

Let $g_1, g_2, ..., g_n$ be n two-dimensional gel electrophoresis images (called hereafter **gels**) and let k be the number of classes to form. k gels are selected (called the **starting gels**) of which each one is assumed to belong to a different class. Thus k classes will be defined, each one with one representative gel. The two steps to be carried out are the following :

1. **Heuristic search.** Insert each of the $n - k$ gels not yet selected into one of the k classes, then give a conceptual description of each class.

2. **Iterative phase.** Select one gel per class, in order to form k new classes, each one with one representative gel, then repeat step 1.

This process is repeated until the classifications converge, i.e. until no better classification can be obtained. The best classification and the descriptions of the corresponding classes form the result.

For each class, the heuristic search step may be viewed as an application of the general data-driven principle given by Mitchell [Mit79]. This learning process corresponds to a two-directional search of the rule and instantiation space, whereby the class descriptions are successively generalized and specified, until only consistent hypotheses are found. These hypotheses must be specific enough to describe all the observed gels belonging to the class and none of the others, but general enough to have prediction capabilities. This strategy is applied separately to each of the k classes and may be split up into five phases :

1. Construction of the maximally specific description of the starting gel g in relation to the other starting gels g_i;

2. Generalization by searching for maximally general descriptions of the starting gel g in relation to the other starting gels g_i;

3. Class filling by inserting all the gels described by the corresponding descriptions;

4. Specification by determining the maximally specific descriptions of the class which the starting gel g belongs to;

5. New generalization, in order to find general descriptions of the class of g.

This algorithm produces k sets, whose elements are the plausible descriptions of a class. Finally, all possible classifications are constructed by taking only one element from each set. Domain-specific heuristics at this stage allow us to reduce the combinatorial explosion and to evaluate the different classifications. The best one is considered to be the result of the heuristic search step.

3   Results

This algorithm has been tested on a set of twelve two-dimensional gel electrophoresis images. The samples used are from liver biopsies of twelve rats, six of them enjoying good health and six suffering from cirrhosis. The number of proteins on each gel varies between 1600 and 2600. The automatic comparison of the gels shows in fact that there are 7651 different spots on the twelve images.

By a way of comparison, the twelve gels have also been classified by a standard clustering analysis algorithm. Eight gels out of the twelve have been classified correctly. The four remaining gels all show a degree of noise higher than average. One gel results from a more concentrated biochemical sample than the others. It is therefore darker and contains more spots. The three others underwent major distortions from different biochemical origins. Only the images of the eight two-dimensional gel electrophoresis produced in ideal experimental conditions could be classified by the numerical taxonomy algorithm. This technique is therefore too sensitive to noise.

The **heuristic clustering** algorithm, for its part, has correctly classified all of the twelve gels, separating the six normal gels from the six cirrhosis ones. Moreover it has also displayed five characteristic spots out of the 7651 spots contained in the gels (fig. 2). From these five spots rules are formulated to detect cirrhosis and are integrated into the expert system.

4   Conclusion

This approach has proved that the lack of human expertise during expert system building may be, in certain cases, compensated in an effective way by an automatic learning system. In particular, the automatic determination of characteristic elements in classes and subclasses of medical images opens new horizons, where future research is very promising.

References

[App87]   R. D. Appel. MELANIE: Un système d'analyse et d'interprétation automatique d'images de gels d'électrophorèse bidimensionnelle; Systèmes experts et apprentissage automatique. Phd thesis, University of Geneva, le concept moderne Editions, Geneva, 1987.

[Fun87]   M. Funk. MELANIE: Un système d'analyse et d'interprétation automatique d'images de gels d'électrophorèse bidimensionnelle; Traitement de l'image et systèmes experts. Phd thesis, University of Geneva, le concept moderne Editions, Geneva, 1987.

[MS83]   R.S. Michalski et R.E. Stepp. Learning from observation: conceptual clustering. In Michalski, Carbonell et Mitchell, éditeurs, *Machine learning*, Tioga Press, Palo Alto, 1983.

[Mit79]   T.M. Mitchell. An analysis of generalization as a search problem. In IJCAI 6, pages 577-582, 1979.

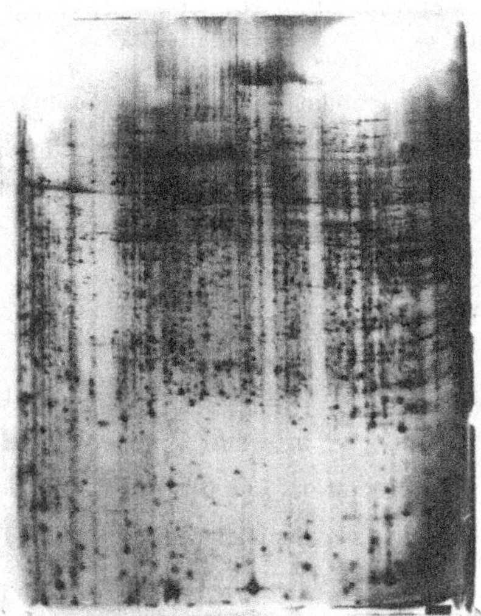

Figure 1: a 2D-gel electrophoresis of human liver biopsy

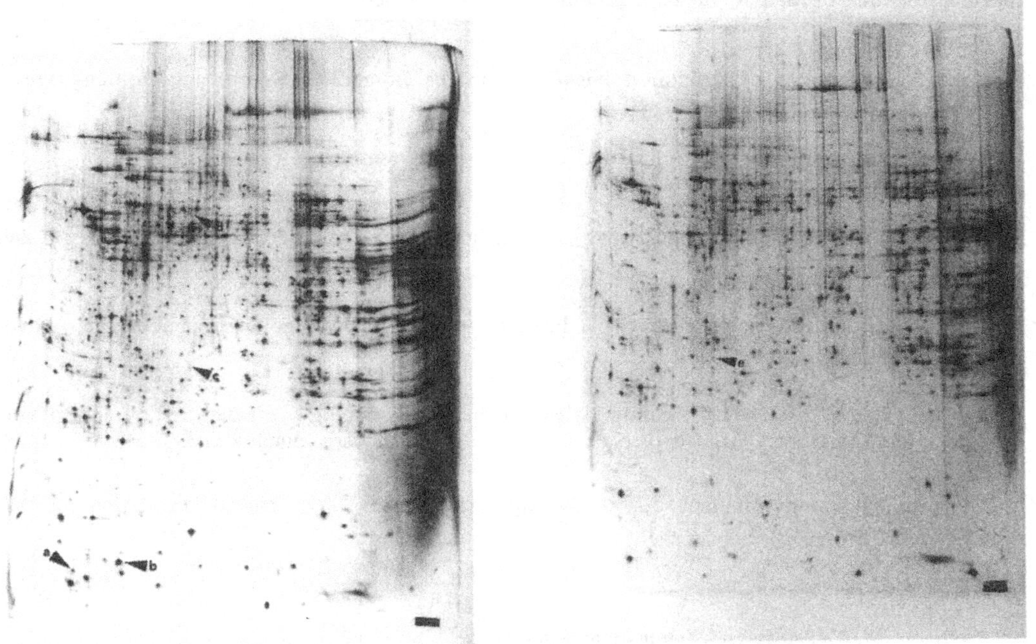

Figure 2: The characteristic spots of a cirrhotic (a,b,c,d) and a normal (e) liver

Knowledge Representation for Cooperative Medical Systems

A.L. Rector
Department of Computer Science
University of Manchester
Manchester   M13 9PL
061-273-7121 ext 5417/5550
email: rector@uk.ac.man.cs.ux

1   Introduction

This paper describes the rationale for the structured portion of a medical knowledge representation system which supports browsing, mixed initiative advice, and multi-level reasoning, the IMMEDIATE Representation Language (IRL). The system has been developed in the course a programme to develop decision support and information systems for general practitioners based on intelligent medical record systems. Although various inheritance networks in frame-like formalisms are popular for medical applications, there are few principled accounts of their functions in medical systems. Most existing systems omit important features needed to manage general medical knowledge.

There are five functions of structured knowledge in the Immediate Representation Language:

*   To deal with defaults, exceptions and common sense reasoning -

*   To guide the search for solutions in methods such as 'heuristic classification' and 'cooperative search';

*   To provide a systematic means of linking surface knowledge with underlying causal representations.

*   To provide a means of dealing with medical terminology efficiently by classifying compound terms such as 'fracture or femur' or pneumocystis pneumonia'.

*   To provide a environment for knowledge engineering and parital validation of the knowledge base.

2   Taxonomies, Defaults and Intensional Knowledge

Medicine's classification systems, or 'taxonomies' provide an essential means of organising and simplifying medical knowledge.

Consider the statements 'All birds fly' and 'Most birds except penguins and ostriches fly'.
The figure below gives a translation of these statements into predicate calculus and a inheritance network notation based on the IMMEDIATE Representation Language.

<u>predicate calculus:</u>   <u>Inheritance Network:</u>

  All birds fly.   Most birds except Penguins and ostriches
                                           fly.

for all X (bird X ==> flies X)   bird: can fly=true

                           penguin: can fly=false

                                   ostrich: can fly=false

The predicate calculus has straightforward translations for what can be said <u>always</u> to be true while the inheritance network has straightforward translations for what can be said <u>usually</u> to be true. The predicate calculus is <u>extensional,</u> that is the truth or falsity of any statement depends solely on the sets of objects which satisfy each predicate. By constrast, systems of common sense or default reasoning are examples of <u>intensional</u> systems; they express what is typically understood or intended to be true. Most medical statements are intensional statements about what is typically true, e.g. 'most cancers metastasize'.

Investigation of the relation between default reasoning and formal logic has revealed the key role played by inheritance networks. Touretzky (1986) has formalised inheritance within semantic networks and proposed a well defined semantics. In previous work the author has extended Touretzky's work to provide criteria for inheritance of binary as well as unary relations, to provide a computationally efficient means of detecting ambiguities, and to extend the concept of inheritance from the primary objects in the system to the atrribute links which connect them {Rector 1986; 1987].

A formal system default theories based on logic was developed by Reiter [1980], but the only computationally tractable subset which he could then identify was too weak to be useful [Etherington & Reiter, 1983]. It has recently become apparent that unambiguous inheritance networks as described by Touretzky can be mapped directly onto a more powerful subset of logical default theories, and that this mapping almost certainly defines the most powerful computationally tractable subset of such theories [Froidveaux 1986; Etherington 1987].

3   Heuristic Search

In a recent paper in which he analyses most of the major medical expert systems, Clancey [1985] claims that 'causal process classification' is the predominate method used in medical diagnostic systems. Figure 1 illustrates the basic pattern. Observed patient data are 'abstracted' up a classification system until they form an understandable unit which triggers consideration of a class of diseases. The diagnosis is then refined from the original category to a specific diagnosis.

The inheritance network is used to summarise the total range of possible causes and therefore reduce the search space. A number of other systems which do not conform completely to the causal process model such as INTERNIST/CADUCEUS [Pople 1982] and our own work on a cooperative search [Rector 1985] also use the structure of the inheritance network to control the search strategy.

Figure 2

Causal Process Classifiaction

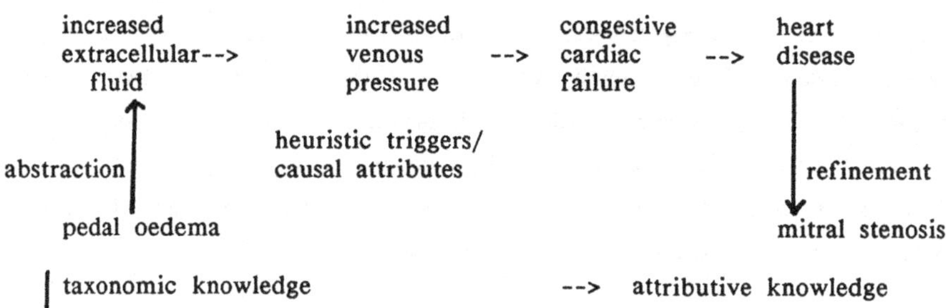

The heart of Clancey's claim is that there are two distinct types of knowledge used in the classification process: taxonomic knowledge which connects instances to subclasses and subclasses to classes, and attributive knowledge which links manifestations and causes directly across taxonomies. (Clancey uses the terms 'hierarchical' and 'non-hierarchical'.) A review of the psychological literature on medical decision making [Patel & Groen 1986; Brooke, Rector & Sheldon 1983; Gale and Marsden 1985] provides additional confirming evidence.

Much of the formal work on inheritance systems eliminates the distinction between attributes and objects, at least for purposes of analysis [Fahlman et al, 1981, Touretzky 1986; Reiter 1987; Attardi 1986]. Clancey's claim is a strong argument for retaining the distinction in practical systems.

<u>4</u>   **Structured Meta-knowledge, Contexts and Levels of Explanation**

Statements about other statements, or meta knowledge, are necessary in order to express contexts and describe the underlying reasoning. An example is shown below in figure 3.

'beta blocking drugs--are contraindicated in--> asthma

                     because

             may cause------> constriction of the airways <——————

             because

             inhibit-->beta adrenergic response >—inhibts———————

simple meta statements such as the above can be considered as 'meta attributes' and summarised using default values in an inheritance network. For example.

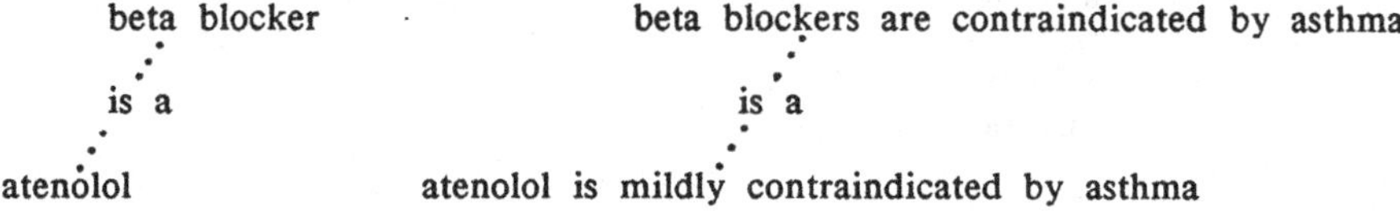

5   Prototypes and terminologic knowledge

New medical terms are often constructed by adding a new anatomic or etiologic descriptor to an existing term, e.g. 'pneumocystis pneumonia', 'dislocation of the proximal interphalangeal joint', or 'e coli septicaemia'. 'Pneumocystis pneumonia' is said to be the 'prototypes' for the attribution 'pneumocystis causes pneumonia'. Prototypes are an example of what Brachman [1981] calls Terminologic knowledge, i.e. knowledge derived from the form or definition of terms rather than from facts about the world.

IRL uses prototypes as a general mechanism for managing definitions and special cases. For example, details of drug treatment depend not just on the drug, but on the disease and patient being treated. Consider the statement 'urinary tract infection may be treated by amoxycillin'. As shown below each node connected by an attribute link generates a prototype. 'Amoxycillin <u>which is</u> used for urinary tract infection contains the specific information on dosages and method of treatment of urinary tract infections by amoxcyllin; 'urinary tract infection <u>which is</u> treated by amoxycillin describes the charactertistics of a urinary tract infection suitable for such treatment'.

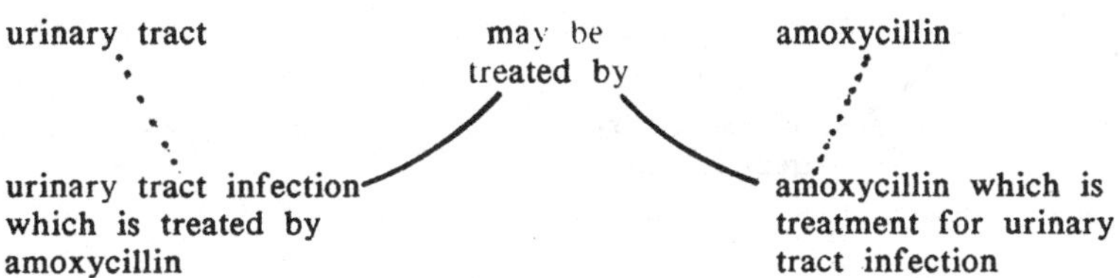

The dosage of drugs is normally less for children than for adults. An attribute 'special case' is used to link a disease or drug with the condition which must be applied, as shown in the example in figure 3. In combining prototypes the subprototype operator <u>and which is</u> is commutative, associative and idempotent; for example: <u>Obj which is C and which is B and which is C and which is A</u> is equivalent to <u>Obj which is A and which is B and which is C.</u> Every prototype has a cannonical form in which every qualifier in the list appears exactly once and the qualifiers are in alphabetical order. Any two prototypes with the same cannonical form are identical.

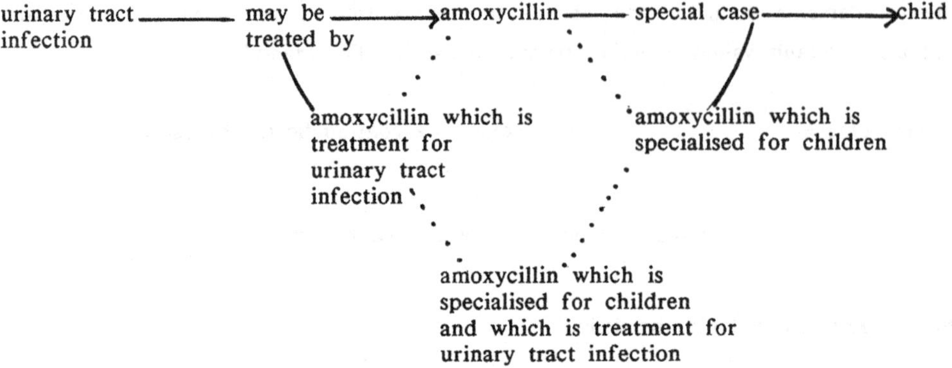

6   Observations and extensional attributions

All of the attribute links discussed to this point have been 'intensional', that is they have been general statements about what was true by definition or default. The prototype of an intensional attribute link may be thought of as the subset of the objects which satisfy the general principle represented.   Observations of particular objects in the world are 'extensional'; they mean that a particular object is an instance of the extension of the corresponding attribute.   For example:

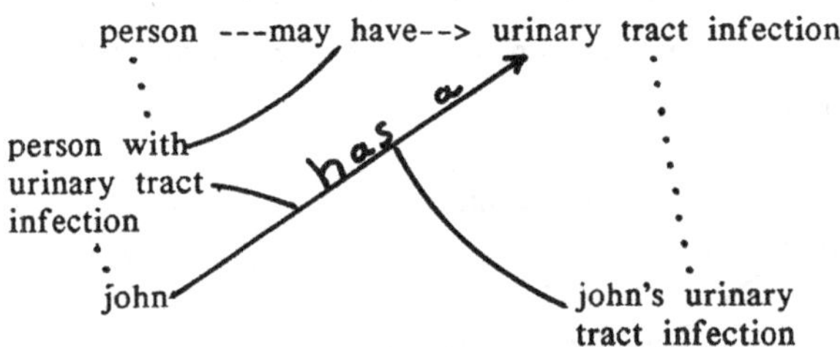

Because 'john is an instance rather than a category, the statement 'john has a urinary tract infection' is equivalent to: 'john is a person with urinary tract infection'.

The second prototype is the instance of urinary tract infection which affects john.   The statements that 'john's urinary tract infection is treated with amoxycillin' would then be represented below.

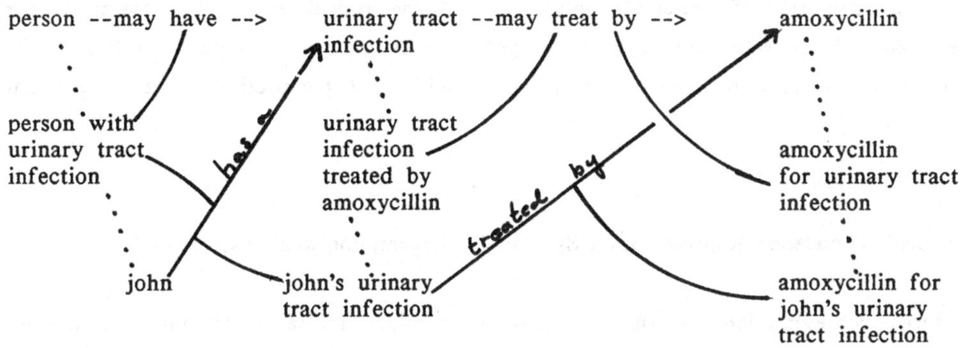

<u>john</u>, <u>john's urinary tract infection</u>, and <u>amoxycillin for john's urinary tract infection</u> are all instances. Each extensional use of an attribute links an instance with a category and creates a further instance.

This representation gives an explicit meaning to each of the linguistic usages for making statements about objects, e.g. 'John has a urinary tract infection', 'John's urinary tract infection is severe', 'The dose of amoxycillin for John's urinary tract infection is 500mg', etc.

7   Temporal statements and the medical record

The central focus of the IMMEDIATE project is the medical record, in which events are recorded in a temporal sequence. Adding time to the above structures is straightforward. Rather than a simple instance, <u>john</u>, we have a hierarchy of instances such as:

```
john at t-2          john urinary tract infection at t-2
   john at t-1          john's urinary tract infection at t-1
      john now             john's urinary tract infection now
```

If it is assumed that all conditions persisted until explicitly stopped, then the same rules of inheritance would apply. One additional rule to account for the expected transition of conditions over time. Each condition has an attribute <u>expected duration</u>. If the expected duration for a condition has been exceeded, an 'exception' is placed in the hierarchy to block further inheritance of the condition.

In general the medical record consists of a series of dated statements concerning patients conditions, the properties of patients' conditions, and the properties of the treatments for those conditions. These statements can be mapped directly onto dated objects in IRL. The object oriented view may be taken merely as an additional perspective on the relational structure.

8   A Medical Knowledge Representation System: Implementation and Discussion

The structured representation described is not a complete inference system. Statements containing existential quantifiers or more than one universal quantifier, such as 'if two drugs have the same effect, then they interact', cannot be represented. Rather the structure provides a skeleton which supports the knowledge engineer in organising general the mass of facts into principles and special cases. The concept of ptototypes described here provides a cleaner solutions to the problems addressed by the system of 'links' and 'usages' previously addressed by the author [Rector 1986, 1987]. Only those cases which are referred to by other objects and which would otherwise be ambiguous in some way need be dealt with explicitly. Once a special case is identified, the system guarantees that the knowledge engineer will be prompted to consider it when dealing with each specialisation of the original concept.

The structure provides a consistent and well defined semantics for default reasoning within a multiple inheritance network and supports the distinction between taxonomic and attributive statements needed by paradigms such as heuristic classification and cooperative search. We expect to be able to allow users to tailor the system to their own needs since the rigid structure means that the effects of their changes will be circumscribed and well defined. The structure allows the medical record to be viewed either as instances of the objects in the knowledge representation system or as a relational data base.

The representation scheme described has developed out of work on medical record and browsing systems [Rector 1986, 1987], and the original IMMEDIATE Representation Language remains the basis of the current applications developments. Two new implementations are underway. One is an implementation of as much of the system as possible in PROPS2 [Fox, et al, 1985]. The PROPS2 implementation provides a clean prototype of the basic axiomatisation, but is limted by its parsing scheme and computational overheads. The second is a complete redevelopment in PROLOG of the IRL system starting form a specification based on a more rigorous axiomatisation.

References

Brachman RJ (1983). What IS-A is and Isn't: an analysis of taxonomic links in
    semantic networks. Computer 16: 30-36
Brachman RJ and Levesque HJ (1984). The tractability of subsumption in frame-based
    description languages, in Proc AAAI-84 34-037
Clancey WJ (1985). Heuristic Classifiction. Artificial Intelligence 27 289-350
Gale J and Marsden P (1985). Diagnosis; process not product, in MG Sheldon JB Brooke
    and AL Rector (eds) Decision Making in General Practice Macmillan Press Ltd,
    London. pp 59-91.
Etherington DW (1987). Formalizing nonmonotonic reasoning systems. Artificial Intelligence
    31: 41-85.
Etherington DW and Reiter R (1983). On inheritance hierarchies with exceptions, in
    Proceedings of AAI-83 pp 104-108.
Fahlman SE, Touretzky DS & van Roggen W (1981). Cancellation in a Parallel Semantic
    Network, in Proceedings of IJCAI-81 William Kaufman Inc, Los Altos, California pp
    257-263.
Fox J, Duncan T, Frost D, and Preston N (1986). The PROPS2 Primer internal report.
    Imperial Cancer Research Fund, Lincoln's Inn Fields, London.
Froidevaux C (1986). Taxonomic default theory, in Proceedings of ECAI-86. pp 123-129
Patel VL and Groen GJ (1986). Knowlege based solution strategies in medical reasoning.
    Cognitive Science 10: 91-116
Pople HE (1982). Heuristic methods for imposing structure on ill-structured problems. in P
    Szolovitis (ed) Artificial Intelligence in Medicine AAAS Symposium 51, Westview Press
    Inc. Boulder Colorado.
Rector, AL (1987). Defaults, exceptions and ambiguity in a medical knowledge representation
    system. Medical informatics 11: 296-306.
Rector, AL (1986). Defaults, exceptions and ambiguity in a medical knowledge representation
    system, in Proceedings of ECAI-86. The 7th European Conference on Artificial
    Intelligence vol II pp 177-182.
Rector, AL (1985). The knowledge based medical record - IMMEDIATE-I - a basis for
    clinical decision support in general practice. in Proceedings of the International
    Conference on Artificial Intelligence in Medicine, Pavia, Italy, September 1985. North
    Holland and Press. pp 37-49.
Reiter R (1980). A logic for default reasoning. Artificial Intelligence 13: 81-132.
Touretzky DS. (1984) The Mathematics of Inheritance Systems. unpublished PhD thesis,
    Computer Science Department, Carnegie-Mellon University, Pittsburg, Pennsylvania 15213,
    USA.

A REPRESENTATION OF TIME FOR MEDICAL EXPERT SYSTEMS

Ian Hamlet and Jim Hunter

Artificial Intelligence in Medicine Group
School of Engineering and Applied Sciences
University of Sussex
Brighton, BN1 9QT

ABSTRACT

Despite the fact that reasoning over time is clearly of great importance in diagnosis, prognosis and treatment planning, few medical expert systems have attempted to represent this facet of medical knowledge. This paper describes how a "time specialist" is being developed to handle the temporal aspects of problem-solving, and in particular, how it might deal with uncertainty and ambiguity in temporal relationships by using a form of Truth Maintenance.

Keywords: expert systems, temporal reasoning, truth maintenance.

1. INTRODUCTION

Doctors do not try to make a complete diagnosis and prognosis based only upon the information obtaining at a single point in time. They understand how diseases progress and how the various findings are temporally related to these processes; in considering a given patient, they try to establish a chronology of the various symptoms, signs and test results. However there may be ambiguity in the temporal relationships, possibly because the patient's memory is not perfect, or perhaps because adequate records have not been kept. Consider the following account, which we will analyse later:

> *"The palpitations started when I stopped taking the tablets - or perhaps just before. I didn't feel any chest pain while taking the tablets. But I remember that I did feel some chest pain at the same time as the palpitations. Yes, that's right, I stopped taking the tablets, and the chest pain started at once - or perhaps a little later."*

Medical knowledge (at various levels) is brought to bear on the findings to produce a number of candidate "explanations" which adequately match the findings and their temporal relationships. After the candidates have been established, the clinician has the opportunity to gather more diagnostic information to aid differential diagnosis. In particular, when test results become available, they serve to further differentiate candidate solutions; however it should be noted that this new information relates to the time when the test was performed, not to the time when the results are received.

By using temporal knowledge about disease processes and the history of a given patient, the clinician is often able to make predictions about the future progression of the disease, and how the patient may react to differing forms of intervention at different times.

Long [1] described how extra diagnostic constraints can be ascertained by augmenting an existing causal knowledge base with a simple representation of duration; he showed how useful even simple temporal relationships can be in aiding diagnosis. VM [2] also contained explicit temporal knowledge within its knowledge base. However, in most existing medical expert systems, useful diagnostic and therapeutic information is almost certainly being ignored because temporal knowledge is not represented. For a survey of previous approaches to temporal reasoning in medical expert systems, see [3,4].

2. A BASIC REPRESENTATION FOR TEMPORAL KNOWLEDGE

Our approach is based on that of Allen [5,6,7] who developed a representation and algebra based upon temporally related intervals. Intervals are continuous finite pieces of time over which some attribute of interest holds. Thus an interval might represent a consultation, a period of elevated blood pressure, a period of tachycardia, an asthmatic attack, etc. Allen prefers the use of intervals to points as the primitive temporal unit, because reasoning with points often produces counter-intuitive results.

2.1. Representation

Knowledge about the relationships between intervals is maintained as a graph [6]. Intervals are represented as nodes in the graph; each arc between a pair of nodes represents the set of possible temporal relationships between the two intervals. The primitive temporal relationships which may hold between two intervals are shown in Figure 1.

Relation	Symbol	Inverse	Example
X equal Y	=	=	XXX YYY
X before Y	<	>	XXX YYY
X meets Y	m	mi	XXXYYY
X overlaps Y	o	oi	XXX YYY
X during Y	d	di	XXX YYYYY
X starts Y	s	si	XXX YYYYY
X finishes Y	f	fi	XXX YYYYY

Figure 1.   Allen's 13 Primitive Temporal Relations

It is important to note that these relationships are *qualitative* in nature; there is no need to know the precise times at which intervals start or finish (although this information can be useful if available). As we have said, the set of relationships on an arc represents the possible relationships which may exist between the two intervals; this allows us to express any desired degree of uncertainty about our temporal knowledge. For example, if two intervals, A and B, are known to be disjoint (i.e. not to overlap in any way) then this can be represented as

A {< m mi >} B

However, it should be noted that the relationships on a given arc are mutually exclusive – one, and only one, of them can be the relationship which actually holds in the real world.

In the account given above, if T represents the interval during which tablets were taken, and P and C are respectively the intervals during which palpitations and chest pain were experienced, then we have for the first three statements:

    T {m o} P
    C {< m mi >} T
    C {= o oi d di s si f fi} P

If nothing is known or nothing can be deduced about the relationship between a pair of intervals, then the arc is labelled with the "unknown-set" which contains all thirteen of the basic relationships:

    A {= < > m mi o oi d di s si f fi} B

2.2. Propagation of Temporal Relationships

Initially, all arcs contain the "unknown set"; however as more information is acquired, these sets get smaller. As the constraints are applied, their effects must be propagated to more remote parts of the graph, to maintain a consistent temporal picture. Thus the relationship set on a given arc is constrained both by direct input from the external problem-solver, and by propagation from other arcs.

Propagation is therefore based upon building the complete graph for the asserted interval-nodes and occurs when an arc is created or its set of relationships updated. The database representing the graph is searched for other node-pairs which have one node in common with the node pair whose relationship set has been changed; the two pairs of nodes then form a three node chain, and by using a temporal algebra and the relationship-sets for each node-pair, the relationship-set between the two end node elements is computed. Thus if the arc AB is updated then the propagation algorithm considers xA with AB to update xB for all x; likewise AB with By to update Ay for all y. Each new relationship-set is queued for assertion into the database.

Before a relationship-set is stored in the database, a check is made to see if there already exists an arc for that node-pair. If the new relationship-set is a superset of, or equal to the existing one then no action is taken, otherwise the intersection of the two is computed and this value is stored in the database; remember that the relationships form a disjunctive set. The algorithm now cycles and tries to propagate this new value through the graph.

The propagate-update cycle continues until no more new information can be derived. If at any time during propagation an arc's relationship-set becomes empty, the user is informed that there are inconsistencies in the constraints used. The algorithm restricts immediate computation to inference chains of length two (i.e. involving three nodes), while ensuring that the entire graph is eventually updated.

In the previous example, CT and TP together generate:

    C {= < > mi d oi s si f} P

which when intersected with the existing relationship for CP gives:

    C {= d oi s si f} P

When we now reconsider CP and PT (the inverse of TP) we obtain:

    C {mi >} T

which is consistent with the final statement in the patient's account.

One positive aspect of this approach is that it preserves the disjunction of relationships and thus avoids unduly restricting the set of possible relationships until necessary. This is important in planning and scheduling where early over-constraining may be unwise [8].

3. USING THE BASIC REPRESENTATION

We now show how the representation described could be used in a second medical example, by mapping medical entities of interest onto the interval structure.

Within medicine there are attributes whose state or value must be one of a closed, possibly ordered, set. Such attributes also frequently have what can be described as a high degree of persistence; this implies that once an attribute is caused to have a particular value, that value will continue to exist unless anything occurs which would change it.

```
A:

            visit_1                    visit_2
         |---------|                |---------|
      |????---------------------------------------?????????????????|
            BP=normal


         visit_1 {<}    visit_2
         visit_1 {d s}  BP=normal
         visit_2 {d f}  BP=normal

B:

            visit_1                    visit_2
         |---------|                |---------|
      |????----------------|------------------?????????????????|
            BP=normal              BP=high


         visit_1    {= d s f}  BP=normal
         visit_2    {= d s f}  BP=high
         BP=normal  {m}        BP=high

C:

            visit_1                    visit_2
         |---------|                |---------|
      |????----------------|------------------?????????????????|
            BP=normal              BP=high


      |????----------------|----------------------??????????????|
          no Hypertension          Hypertension


                           |???-----------------???|
                             Retinal problems
```

Figure 2.

In Figure 2A, during visit_1, a patient is known to have normal blood pressure (BP): our default expectation ("persistence") is that this will be maintained. At the next visit, if the patient is now found to have a high blood pressure and we suppose blood pressure to have only two qualitative values (normal and high), we may represent this as shown in Figure 2B. Because there are two mutually exclusive values for the same attribute, we can constrain the relationship between the two intervals to be "meets". This is a simplification of the real problem since there are more complex patterns of change that could occur. For example instead of simply changing from normal to high, the pattern of change could have been:

normal -> high -> normal -> high

As there has been a change in the value, then something must have caused the change and the problem-solver must search for possible causes. Attribute/value changes are only caused by events or processes. In this example the problem-solver hypothesises that essential hypertension is the cause of this change. In this case we can assume that as soon as hypertension was present then the high blood pressure was also present. Hypertension has a number of other signs which are usually present, for example the deformation of retinal blood vessels (se Figure 2C).

The possible effects of the hypertension should also have their attribute/values and temporal relationships added to the database. It will then be possible to identify those attributes which might be present during a visit interval; thus retinal problems might or might not be observed duing visit_2 depending on the temporal relationship between the time of onset of hypertension and the time of onset of the retinal problems.

We can use some of the primitive relationships described earlier to provide a simple mechanism for representing causality; at the surface level there seem to be only a few possible relationships as illustrated in Figure 3. Interval A might be one where the blood volume is high, and B an interval where oedema is present.

```
                           A causes B           Relationship

      Type 0          |----------|A          A {=}  B
                       |----------|B

      Type 1          |----|      A           A {s}  B
                       |----------|B

      Type 2          |----------|A          A {fi} B
                             |----|B

      Type 3          |-------|A             A {o}  B
                          |-------|B

Arguably there may be two other types required

      Type 4          |-----------|A          A {di} B
                           |----|B

      Type 5          |----|A                 A {<}  B
                             |-----|B
```

Figure 3. Causality in terms of temporal relations

These examples are intended to show the different ways in which "cause and effect" may be temporally related; they differ in whether there are lead and/or lag times between onsets and cessations.

4. THE EXTENDED TEMPORAL REASONER

Any medical expert system has to be capable of coping with uncertain knowledge and data. The more traditional approach is to try to quantify the uncertainty in some way (e.g. MYCIN's certainty factors) and then to propagate these quantities using an appropriate calculus. Our approach is rather to make alternative assumptions about the state of the world (suitably constrained) and to explore the consequences of these different alternatives. In the context of temporal reasoning, uncertainty will manifest itself in two different ways.

Firstly we may hypothesise the existence of a certain event (e.g an episode of angina, or a course of a certain medication) and investigate how our understanding of the patient's history is affected by the introduction of the assumption that the event actually took place.

Secondly, there may be imprecision in the temporal relationships between two intervals as initially specified, and we may want to explore the assumption of imposing further restrictions. In the first example we might want to see what relationships are possible for CP if we make the assumption that TP is restricted to

$$T \ \{o\} \ P$$

In fact we find that under this assumption:

$$C \ \{d \ oi \ f\} \ P$$

The problem with Allen's original representation is that it maintains only *one* view of the world, and does not allow the problem-solver to compare different hypothetical worlds, both past and future.

4.1. Use of Data Dependencies

Our approach to a solution to this problem is inspired by de Kleer's work on Assumption Based Truth Maintenance Systems (ATMS) [9]. The basic idea is that we identify the alternative assumptions that we make about our model of the temporal world. As discussed above, these assumptions are at two different levels. Firstly there is the assumption that a certain event (interval) exists with a given set of relationships to another given interval. Secondly there is the selection of one relationship within a given disjunctive relationship-set. Propagation through the graph takes place as before, but each derived relationship is labelled with the basic set of assumptions on which it depends; there may be more than one way of deriving a particular relationship - in this case the label consists of a number of underlying assumption sets.

A possible view of the world consists of set of assumptions, called a context. Rather than having just one context, which is implicitly what is happening in the basic temporal reasoner, the problem-solver can create whatever contexts it likes to reflect the particular hypothesis that is under consideration. By comparing the labels on the relationships with a given context, the temporal reasoner can answer questions such as what relationships hold in that context, or whether a given relationship holds.

In propagating these assumption sets we must make sure that certain inconsistencies are not introduced. Firstly, relationships for a given interval are mutually exclusive. For example, we can not simultaneously believe:

$$A \ \{<\} \ B \quad AND \quad A \ \{m\} \ B$$

Recall that

$$A \ \{< \ m\} \ B \quad means \quad A \ \{<\} \ B \quad XOR \quad A \ \{m\} \ B$$

Thus any attempt to label A $\{<\}$ B with an assumption set which includes A $\{m\}$ B must be prevented. Secondly, we must make sure, for any context the problem-solver choses to investigate, that for each arc there is at least one relationship which could hold in that context (since any two intervals must have at least one possible relationship).

4.2. A Problem-Solver's Use of the Extended System

The approach advocated here is one of providing a problem-solver with a sub-component, a sort of time specialist, which is specifically orientated towards dealing with temporal relations. This is similar to the ideas put forward by Kahn [10,11]. In many respects it would augment a conventional problem-solver by allowing it to take advantage of some of the extra constraints that may be derived from explicit temporal relations. The temporal component should not depend upon the type of representation used in the medical knowledge base, as long as the knowledge representation can support this temporal annotation.

The representation is not sensitive to "direction" in time; planning for the future and diagnosis of past events can be approached in a similar way because time is represented only in a relative fashion. For instance a relationship between high blood volume and oedema might be represented as:

high blood volume <==>> oedema

the "<==>>" symbol being used to represent the non-symmetry between cause and effect. In diagnosis, this would mean that if oedema is present, high blood volume is a likely possibility. In prognosis it would mean that given a high blood volume, oedema is likely to follow.

The time specialist would be responsible for maintaining the overall temporal relationships between events, processes, findings, and projections. The problem-solver will need to interface with the time specialist via appropriate access and inquiry functions and predicates which would answer questions such as "does X hold/occur during Y ?" and "is X still true ?".

Because the time specialist internally builds up a relative temporal picture there are no problems in allowing an interval to represent the present time. This "now-interval" can be changed to represent the passage of time, and the system already has the past and futures computed. In the second example, the visit intervals could be used as the "now-intervals" and the time specialist would easily be able to determine what should still be true or present, what should have ceased and what should have started. The problem-solver may validate these details to see if the assumptions that caused them to be believed do receive confirmation. If the expected results do not occur, the problem-solver must determine if the assumption(s) on which the expectation is founded are at fault and if so, reject any hypothesis based on them.

4.3. Implementation Status

We have implemented Allen's temporal reasoner and the ATMS separately in POP11, and have interfaced them in various ways to test the ideas discussed above. In order to improve the efficiency of the system, we intend to proceed to a versions which more fully integrate the two. We have also yet to develop a convenient interface for problem-solvers.

5. DIFFICULTIES

The first problem is one of size: the time specialist at the moment maintains a complete network of all nodes, so the addition of a new node into the system requires the computation of arcs to each existing node. Analysis shows the propagation algorithm to be polynomial in time with respect to the number of nodes [12]. The algorithm, although sound, is incomplete. The incompleteness arises from the fact that it only performs three-ary propagation and as such can only detect inconsistencies in any three-node subgraphs. Inconsistent labellings can persist undetected in four-node subgraphs, and the price for completeness in an n-node network is the solution of an NP-hard problem. There are simple mechanisms for checking the validity of a small subgraph by a simple backtracking search to find single arc-labellings. Although the incompleteness is a deficiency with the algebra, it has not actually proved to be a problem in practice.

One approach to the size/complexity problem is to introduce more structure into the temporal database. When we talk about events we can organise them hierarchically to minimise the number of explicit arcs that have to be kept between nodes. In the algebra, concise hierarchies may be built up around the "during" relationship. This approach works well until the pure hierarchical structure starts to break down, and the advantages then start to diminish. More work needs to be done in looking at how the database can be structured, possibly automatically, to minimise the amount of propagation that is necessary.

6. CONCLUSIONS

In this paper we have tried to bring together the elements for a powerful and flexible time specialist which would manage the temporal aspects of medical knowledge and data for a higher level problem-solver. It would allow uncertainty to be handled by enabling the problem-solver to compare expectations in different hypothetical worlds, maintained concurrently; the problem-solver could switch between and modify these worlds as required with little computational overhead. We also feel that this approach may help to provide a unifying approach to diagnosis, prognosis and patient management in that it allows them all to be viewed on the same time line stretching continuously from past to future.

ACKNOWLEDGEMENTS

We gratefully thank all of the members of the AI in Medicine Group for helpful discussions, in particular Nick Gotts for his comments on early drafts of this paper, and Roger Sinnhuber for his implementation of the ATMS. One of us (IH) is supported on an SERC studentship.

References

1. W.J. Long, "Reasoning about State from Causation and Time in a Medical Domain," *AAAI-83*, pp. 251-254, 1983.

2. L.M. Fagan, "VM: Representing Time-Dependent Relations in a Medical Setting," PhD Thesis, Stanford University, Dept. of Computer Science, 1980.

3. R.K. Sinnhuber and J.R.W. Hunter, "Temporal Information for Medical Expert Systems: A Review of Recent Work," Report AIMG-6, Artificial Intelligence in Medicine Group, University of Sussex, 1984.

4. J.R.W. Hunter, "Artificial Intelligence in Medicine: A Tutorial Survey - Part 2: Comparisons," *Biomedical Measurement, Informatics and Control*, 1987.

5. J.F. Allen, "An Interval-Based Representation of Temporal Knowledge," *IJCAI-81*, 1981.

6. J.F. Allen, "Maintaining Knowledge about Temporal Intervals," *Communications of the ACM*, vol. 26, no. 11, 1983.

7. J.F. Allen and H.A. Kautz, "A Model of Naive Temporal Reasoning," in *Formal Theories of the Commonsense World*, ed. J.R. Hobbs and P.C. Moore, pp. 251-268, Ablex, Norwood, New Jersey, 1985.

8. J.F. Allen and J.A. Koomen, "Planning Using a Temporal World Model," *IJCAI-83*, pp. 741-747, 1983.

9. J. De Kleer, "An Assumption-Based TMS," *Artificial Intelligence*, vol. 28, no. 2, pp. 127-162, 1986.

10. K.M. Kahn, "Mechanisation of Temporal Knowledge," M.Sc. Thesis, MIT/LCS/TR-155, 1975.

11. K.M. Kahn and G.A. Gorry, "Mechanising Temporal Knowledge," *Artificial Intelligence*, vol. 9, pp. 87-108, 1977.

12. M. Vilain and H. Kautz, "Constraint Propagation Algorithms for Temporal Reasoning," *AAAI-86*, vol. 1, pp. 377-382, 1986.

Management of Uncertainty

TOULMED, an inference engine which deals
with imprecise and uncertain aspects
of medical knowledge

J.C. BUISSON, H. FARRENY, H. PRADE

Laboratoire Langages et Systèmes Informatiques

ENSEEIHT et Université Paul Sabatier, Toulouse

M.C. TURNIN, J.P. TAUBER, F. BAYARD

Service d'endocrinologie

CHU Rangueil, Toulouse

Abstract: TOULMED is an inference engine, capable of accomodating imprecision and uncertainty, which is developed and experimented in medical applications. The illustrative examples in this paper are taken from the application to the treatment of diabetes. The paper emphasizes the importance of the management of imprecision and uncertainty in order to take into account all the nuances of the medical knowledge. More particularly, a comparative example is provided where the same medical knowledge is processed with and without a treatment of imprecision and uncertainty (which is dealt with in the mathematical framework of possibility theory).

I-Introduction

1- diabetologists concern

DIABETO is an expert system, built on the inference engine TOULMED in diabetology. It has been developped to supply diabetics and their private doctors with information concerning their illness, allowing them to have a better control of it.

Diabetes is a chronic illness, leading to numerous complications which at times are very incapacitating. This requires from the diabetic patient to play an active role in their therapy and to modify their personal hygiene and life style as a mean to prevent these complications.

It is thus very important that the diabetic patient and his private doctor be informed and given advise on the ways to handle and control the disease at home on a day to day basis (home serum glucose monitoring - diet - physical activity - insulin dose adjustement etc...).

The DIABETO expert system, easily accessible through the French Minitel network, allows to educate the diabetic patient by individualizing the necessary advises concerning changes in his life style and personal hygiene. This will facilitate the work of the treating physician without replacing him, since no actual drug prescription are done by the system. Such a system will also allow the execution of epidemiologic studies that might contribute to our present knowledge of diabetic self monitoring and care.

We already tried a preliminary format of the system in a hospital environment, and an experiment with the diabetics using the system at home with the help of their private doctors is planned. A validation phase will also be necessary: this will be done through other French diabetology centers.

In addition to its medical benefit, the program should be able to keep constant the attention of the user. The input of data has been simplified, and the results are displayed in many small sets. In other respects, a link has been established between the expert system and a documentary data base <6>, in order to provide technical information on precise subjects. The expert system itself build the queries (in terms of keywords) addressed to the documentary system, and the pages of information are then adapted to the personal situation of the patient.

2- the expert system approach

An expert system-based approach in medicine is justified by the granularity of the knowledge to deal with, moreover, the set of medical rules we use was reformulated several times in many cases, although the medical knowledge is well-established. For instance, we had to add and/or modifiy several rules when we have taken into account the imprecision of conditions and the uncertainty of conclusions. These changes were the result of the expert efforts to express their knowledge better, in a more shaded and adequate manner.

Besides, there is an increasing number of medical expert systems, often dealing with uncertainty (e.g. <10>, <11>, <12>), which have reached an advanced state of development, or are in the course of experimentation. However, it does not yet exist a system which actually gives an effective help for general practitionners, and then we have chosen to design and develop a new inference engine, called TOULMED, well adapted to the medical field and which allows us to perform and control different types of inference, taking into account imprecision and uncertainty in a non ad hoc manner. At the moment, TOULMED is not only used in diabetology, but also in anesthesiology and odontology among other medical fields.

3- necessity of dealing with the imprecision and the uncertainty of medical knowledge

At first glance, we might think that the information given by the patient or obtained from medical analyses could be regarded as precise without any major inconvenience, and that the categories which are used in the expression of the medical knowledge could also be considered as having sharp and precisely located boundaries. In fact, that is not really the case. Indeed, it may be important to determine clearly if the result of a medical analysis or if an aspect of the description of a symptom can or cannot be definitely classified in a given category, since according to the category, the conclusions which might be derived subsequently may be very different.

Since the expert system MYCIN <2>, which has been the first to deal with uncertainty), many other numerical models have been proposed. Possibility theory offers a framework for treating imprecision and uncertainty in a rigorous, unified and computationally tractable manner. This paper presents, on illustrative examples, the interest and the main features of the management of both imprecision and uncertainty in TOULMED, based on possibility theory. For a more detailed and technical presentation of the approach, see <4>,<5>.

In section II, a concrete example emphasizes problems encountered in diagnosis when processing with too sharply defined rules and facts. The section III introduces the principles of representation and treatment of fuzzy knowledge.

II- Presentation of a concrete example

The comparison of two examples of dialogues, extracted from real consultations with TOULMED, will illustrate the advantages of taking into account the imprecision and the uncertainty of the medical knowledge.

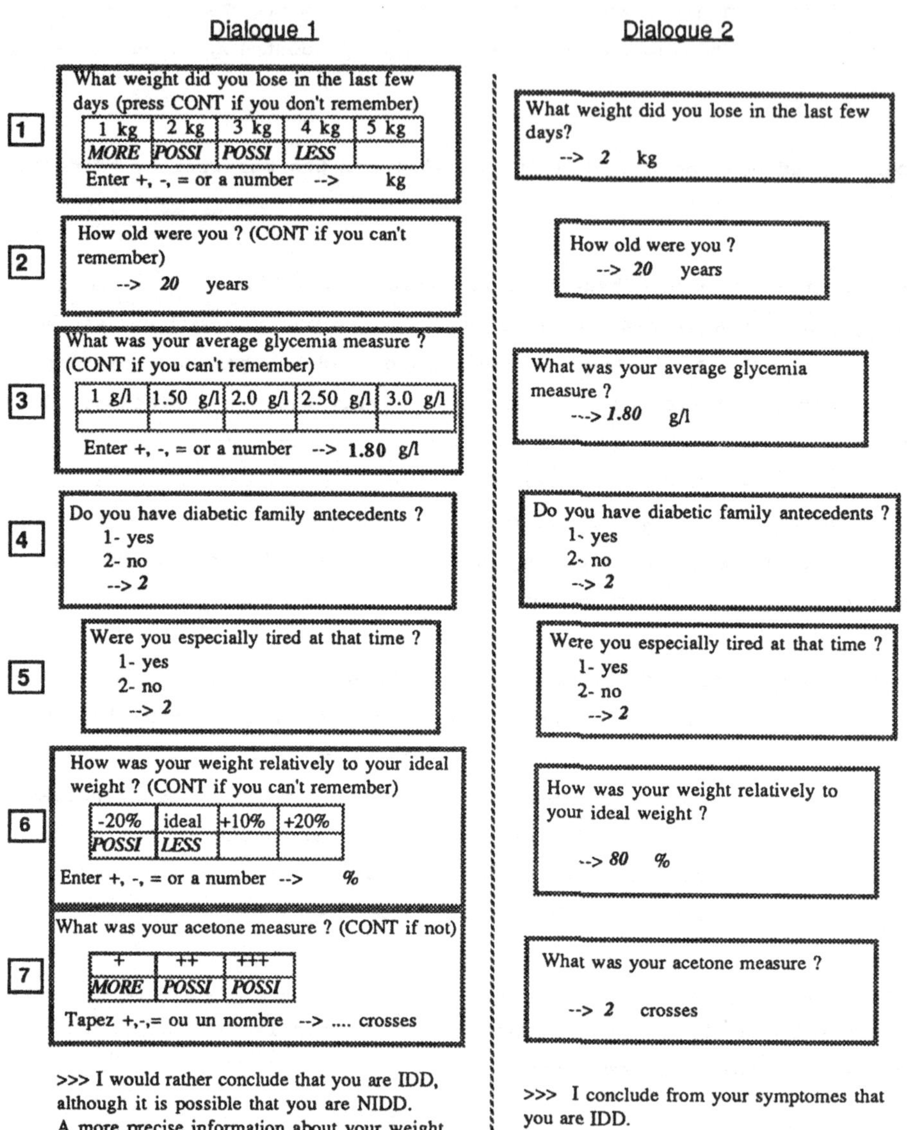

Both dialogues in figure 1 correspond to the processing by TOULMED of the
non-formalized rules of figure 2. These rules, written in natural language,
were formalized in two different manners, both close to the internal codification used by TOULMED.

r1- very strong weight loss --->   certainty of IDD

r2- strong weight loss and less than 30 years old ---> high presumption of IDD

r3- strong weight loss and more than 30 years old and high glycemia measure

       and high acetone measure ---> very high presumption of IDD

r4- strong weight loss and high glycemia measure ---> high presumption of IDD

r5- no strong weight loss and high acetone measure

       and significant tiredness and high glycemia measure

           ---> high presumption of IDD

r6- high acetone measure and high glycemia measure

       and less than 30 years old ---> high presumption of IDD

r7- no strong weight loss and weight higher than ideal weight

       ---> presumption of NIDD

r8- no strong weight loss and family antecedents   --->   presumption of NIDD

r9- no strong weight loss and weight higher than ideal weight

       and no high glycemia measure and no significant tiredness

       and no high acetone measure --> certainty of NIDD

IDD: "insulino-dependent diabetes"; NIDD: "non insulino-dependent diabetes"

Figure 2: the set of medical rules used in the example, as written by the physicians. To be used by TOULMED, those rules have to be encoded first.

In the first formalization, all the categories are delimited by precisely located boundaries, and the different numerical parameters are represented by single numbers, sometimes asked for directly to the user. Moreover, all the rules are supposed to be certain, and the qualifying terms "strong presumption", "very strong presumption" ... of the original rules are ignored. The figure 3 gives two examples of such a formalization, in which all the fuzzy aspects are neglected. The use of these rules by TOULMED leads to the dialogue 2 of the figure 1 (imprecision and uncertainty is taken into account in dialogue 1, not in dialogue 2).

R2- weight loss >= 2 kg   and   age < 30 years   --->   IDD

R7- weight loss < 2 kg   and   relative-weight >= 120%   --->   NIDD

Figure 3: formalization of r2 and r7 in a form close to the rules directly used by TOULMED. They do not take into account the uncertainty of the conclusions ("high presumption",...), and the imprecisely bounded categories ("strong weight loss", ...) are represented by intervals. Examples of facts which describe the patient are: weight loss=2kg, age=45 years, relative weight=110%; the information associated with a parameter is its value, one supposed that the patient knows them with precision and certainty.

In the second codification presented in figure 4, each parameter which concerns the patient is no longer represented by a single number, but with a *possibility distribution* (in short: p.d.) which expresses both imprecision and uncertainty about the knowledge of the associated variable. For instance, the p.d. represented by the 5-tuple (2   3   0.5   0.5   0) expresses that the patient has lost about 2 or 3 kilograms, in any case (certainty degree = 1) a value which does belong to the interval [2-0.5, 2+0.5] (see III-1 for more details and examples about the possibility distributions). At each rule is also attached a value in [0,1] which expresses to what extent the expert is sure that the conclusion is true when all the conditions of the rule are completly satisfied.

R2'- weight-loss rather>= 2 kg   and   age rather< 30 years

--->   high presumption of IDD   (0.7)

R7'- weight-loss rather< 2 kg   and   relative-weight rather>= 120%

--->   presumption of NIDD   (0.4)

Figure 4: formalizations of r2 and r7 which take into account of the uncertainty of the conclusion (0.7 for R2', 0.4 for R7') and which use fuzzy categories ("rather>=", ...). Besides, the knowledge associated with a parameter is a possibility distribution, for instance:
weight-loss : (2 3 0.5 0.5 0) which gives information about the value of weight-loss.

The input of imprecise numerical parameters is done by means of 'grid table' as in questions 1,3,6,7 of dialogue 1. We can see on question 3 that it is possible for the user to give directly the precise value of the parameter, if he or she knows it; therefore that sort of input is more powerful than the usual numerical input of dialogue 2. In other respects, even if the user do not know the exact value of a parameter, he can say something about it in general. This is the role of the grid to collect such information.

By writing POSSI (pressing the "=" key) in front of a value, the user expresses that this value is "possibly" the value of the parameter. By writting MORE (resp. LESS), pressing the "+" (resp. "-") key, he indicates that the real value is rather more (resp. less) than the considered value. If he/she do not know anything about the parameter, he may press the CONT key, while he is obliged to give a precise value in the numerical questions of dialogue 2.

In the example, the advantages of the treatment of imprecision and uncertainty are obvious, since questions are refering to an old period of the patient's life (sometimes several years ago), and since the questions are answer by the patient. Imprecision and uncertainty pervading values are due to the limited capabilities of the patient's memory, and the limited reliability of his measure instruments. For instance, it is difficult to give a value to the weight loss at the beginning of the diabetes (question 1), and the numerical answer of the dialogue 2 (2 kg) is too precise for being fully reliable. However, the imprecise answer might be also the one of a patient who knows his balance well, and who says: "I lost may be 2 or 3 kilograms, but certainly more than one and less than four". The shaded formulation of the conclusion in dialogue 1 is more satisfying than the categorical conclusion of dialogue 2, because the considered case is not very typical with respect to the expert rules r1, r2,... It gives a necessary nuance, because the opposite diagnosis is not excluded.

Besides, a very important improvement offered by the treatment of imprecision and uncertainty appears by looking at the changes of the final conclusion when input information is modified progressively, especially in the near of some boundaries. Figure 5 gives such examples, were all the input data are the same than the input data of the example of figure 1, except for the two parameters *weight-loss* and *acetone*, which are modified.

Dialogues obtained by using the 'crisp' rules (that is to say certain, with non fuzzy categories) in the right part of figure 5 have many deficiencies. Like in the preceding example, the numerical questions often collect arbitrary values (1.5 kg for weight-loss in the case 4, contrastedly with the imprecise input which shows that any value in]0,3[is possible, according to the patient).

Moreover, a real medical danger appears in the conclusions and their formulation. The final conclusions of cases 2,3,4 in the right part of figure 5 make us believe that the situation of the patient is equally obvious in the three cases, without showing the decreasing compatibility of the facts with r9.

Fuzzy rules and facts	'crisp' rules and facts
1) weight-loss : (1 kg 2 kg 3 kg 4 kg 5 kg) *more more more more more* acetone : *CONT* ---> I am completly sure that you are IDD (r1,r2).	**1)** weight-loss = 6 kg acetone = 2 crosses ---> I conclude from your symptoms that you are IDD (r1, r2).
2) weight-loss : (1 kg 2 kg 3 kg 4 kg 5 kg) *less* acetone : *less* + ++ +++ ---> I am completly sure that you are NIDD (r9).	**2)** weight-loss = 0 kg acetone = 0 cros ---> I conclude from your symptoms that you are NIDD (r9).
3) weight loss : (1 kg 2 kg 3 kg 4 kg 5 kg) *possi less* acetone : *less* + ++ +++ ---> It is almost certain that you are NIDD (r9).	**3)** weight-loss = 1 kg acetone = 0 cros ---> I conclude from your symptoms that you are NIDD (r9).
4) weight-loss : (1 kg 2 kg 3 kg 4 kg 5 kg) *possi possi less* acetone : *minus* + ++ +++ ---> It is very likely that you are NDID, but the other possibility is not excluded (r9).	**4)** weight-loss = 1.5 kg acetone = 0 cros ---> I conclude from your symptoms that you are NIDD (r9).
5) amaigr : (1 kg 2 kg 3 kg 4 kg 5 kg) *more possi less* acetone : *less* + ++ +++ ---> Your symptoms are not very clear. It is very possible that your are NIDD, but it is not certain (r2,r9).	**5)** weight-loss = 2 kg acetone = 0 cros ---> I conclude from your symptoms that you are IDD (r2).
6) weight-loss : (1 kg 2 kg 3 kg 4 kg 5 kg) *more possi less* acetone : *possi possi less* + ++ +++ ---> You present very contradictory symptoms. I can't conclude anything (r2,r6,r9).	**6)** weight-loss = 2 kg acetone = 2 crosses ---> You presente both typical IDD symptoms and NIDD symptoms. I can't conclude anything. (r2, r6).

Figure 5: six diagnoses where input data are modified for weight loss and acetone, and where the other parameters keep the values of figure 1. The rules fired are indicated at the end of each conclusion. Note that conclusions can be categorical also with fuzzy rules; note also the evolution of the conclusion in the cases 3,4,5,6.

This rigidity reaches its rupture point when we go from case 4 to the very close case 5, with a complete changing of the conclusion. In the same manner, we go from 4 to 6, or from 5 to 6, from a categorical conclusion to an unsolvable conflict.

At the opposite, the dialogue obtained by using fuzzy facts with fuzzy rules (eventually uncertain, and with imprecise categories) does not exhibit any of the preceeding deficiencies. We can note on cases 1 and 2 that categorical conclusions can also be obtained, even if input information is very imprecise (in the case 1, the answer about acetone expresses a complete lack of information).

On the other hand, answers to cases 2,3,4 show an increasing uncertainty of the diagnosis, which leads to a conflict between the rules r2 and r9 (resp. r2,r6,r9), in the case 5 (resp. case 6). In the case 5, the conflict is judged not very significant, and is solved in favor of a diagnosis of NIDD diabetes (while the crisp rules lead to the opposite diagnosis!). In case 6, a more important conflict is not solved. We shall see in III-1 that a conflict indicates some inconsistency in the set rules+facts.

The absence of a precise diagnosis in a system which uses fuzzy rules may come from the fact that none of the rules have given a significant conclusion, or that a conflict has appeared between the conclusions of several rules, if none prevails enough over the others. In figure 6 are represented the five possible situations from this point of view, and the transitions from a situation to another when a slight changing of the input data occurs.

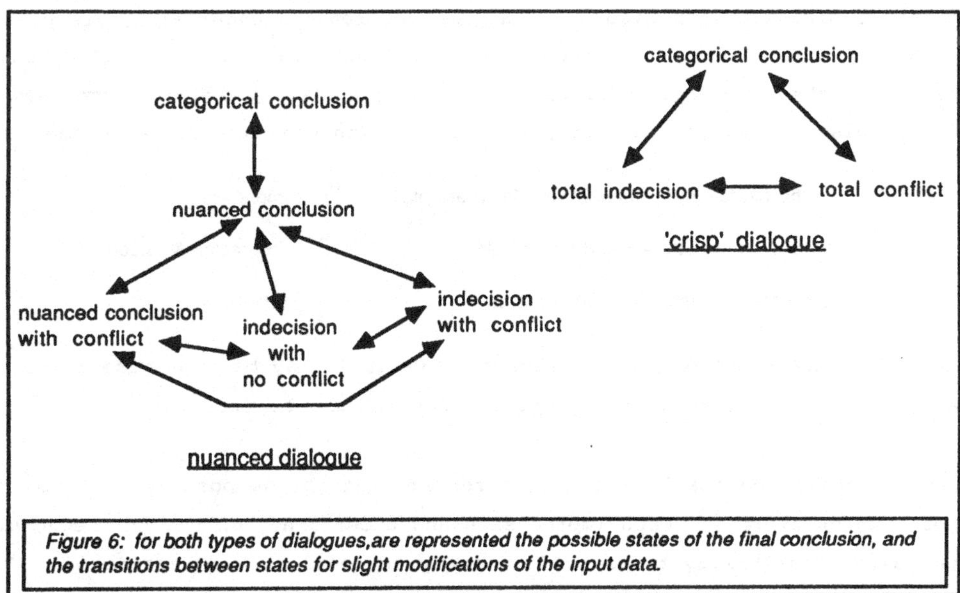

Figure 6: for both types of dialogues,are represented the possible states of the final conclusion, and the transitions between states for slight modifications of the input data.

With a crisp system, the absence of a precise diagnosis may come from a total indecision (none of the rules have been fired) or a total conflict (2 fired rules have given two opposite conclusions). In both cases, the system do not have any qualitative information in order to give a judgment. In figure 6 are represented the three possible situations in the crisp case; it shows how more ˇvaried are the answers and the possible transitions with an expert system which uses fuzzy rules.

III- Principles of the representation and the treatment of imprecision and uncertainty

Without going into all the details of the mathematical formalization, these part presents the main ideas of the representation of fuzzy rules and facts on the one hand, and of their treatment by the inference engine on the other hand. These explanations will be based on the examples of part II.

1- the facts

The medical knowledge, such as the one in figure 1, are referring to many variables: *age*, *glycemia-measure*, *type-of-diabete*,... The meaning of these variables is not ambiguous, and each of them has <u>one</u> precise value for a given patient. What is fuzzy, is *the knowledge* on the real value of the variables; so we can see in the dialogue 1 of figure 1 that *imprecision* and *uncertainty* are present on the variables linked to the input (*weight-loss*, *age*, ...) as well as the variables associated with the conclusion (*type-of-diabetes*). Here are some examples of imprecise and uncertain facts, associated with the same variable *age*.

the patient is between 25 and 28 years old	imprecise
the patient is likely about 30 years old	imprecise & uncertain
the patient is very likely 30 years old	uncertain

We see that the assessment of a value to a variable may be imprecise or not, and that it is the whole resulting proposition which may be uncertain.

In this context, we shall call <u>fact</u> a restriction on the possible values of a variable, which expresses both imprecision and uncertainty of the knowledge of the true value. Possibility theory offers a way to model such a restriction. We associate to each point of the domain of the variable a *possibility degree* belonging to [0,1], which represents the *possibility* that the point is <u>the</u> real value of the variable, according to the represented fact.

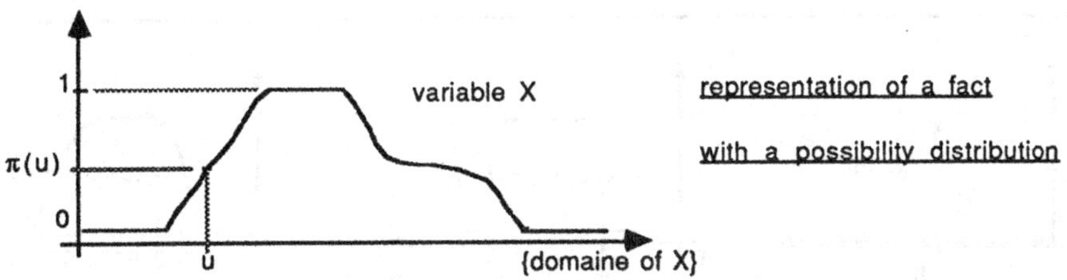

Points of the domain associated with a possibility degree equal to 1 have the maximum "possibility" to be the value of the variable. The points having a possibility degree equal to 0 cannot be the value of the variable in any case; the points associated with a possibility degree between 0 and 1 are as much "possible" as this degree is close to 1. Note that at least one of the points of the domain must have a possibility degree equal to 1 (if not, the domain is not complete). Note also that the domain of the variables may be anything: it is often a continuous portion of the real line, but it can be also a finite set. For instance, the variable *type-of-diabetes* take its value in the set {IDD, NIDD}.

A unexperimented user understands well this language in terms of 'possibilities' (often better than the probability language). It is used in the example of the part II, for the formulation of the conclusion as well as for the input of information; in that manner, the inference engine directly builds up a possibility distribution from the indications given by the user in the grid-table :

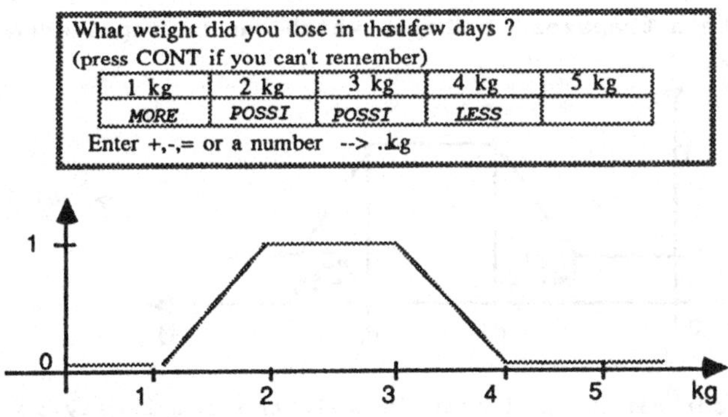

What weight did you lose in the last few days ?
(press CONT if you can't remember)

1 kg	2 kg	3 kg	4 kg	5 kg
MORE	POSSI	POSSI	LESS	

Enter +,-,= or a number --> .kg

Figure 7 gives examples of possibility distributions representing imprecise and/or uncertain facts. The first three ones are representing imprecise but certain facts, since only one subset of the domain is possible {u| $\pi(u)>0$} and since all other points are completely impossible: we are *certain* that the value of the variable is in this subset. On the contrary, the fact represented by $\pi 4$

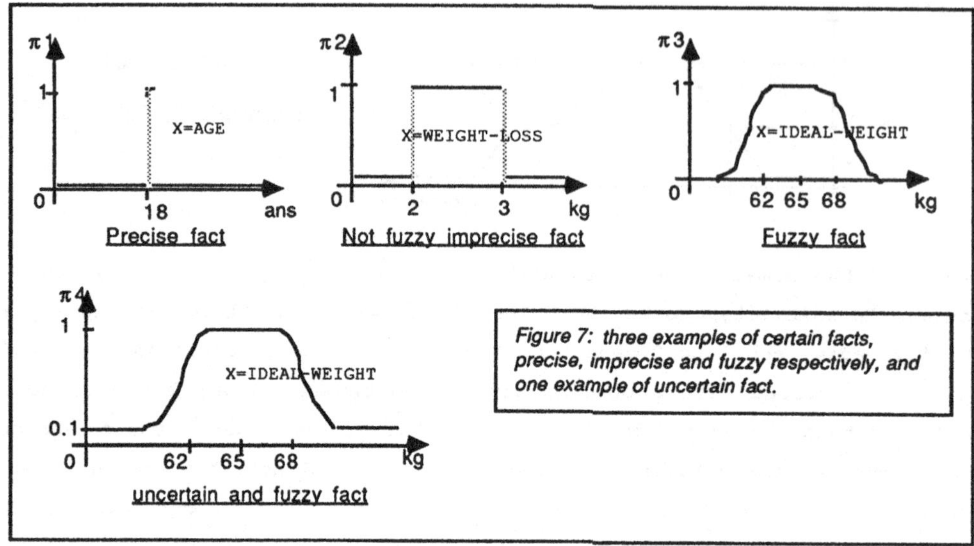

Figure 7: three examples of certain facts, precise, imprecise and fuzzy respectively, and one example of uncertain fact.

is uncertain, since all the points of the domain have a possibility at least equal to 0.1; $\pi 4$ was obtained by adding an undetermination level of 0.1, representing the *uncertainty degree* of the fact. So a possibility distribution appears to be a simple formalism able to represent both imprecision and uncertainty in a unified way.

Given a possibility distribution, it is important to know the points having a possibility equal to 1 (completely possible), and those having a possibility equal to 0 (completely impossible). In practice, TOULMED operates on possibility distributions with a trapezoidal shape, which can be represented by a 5-tuple.

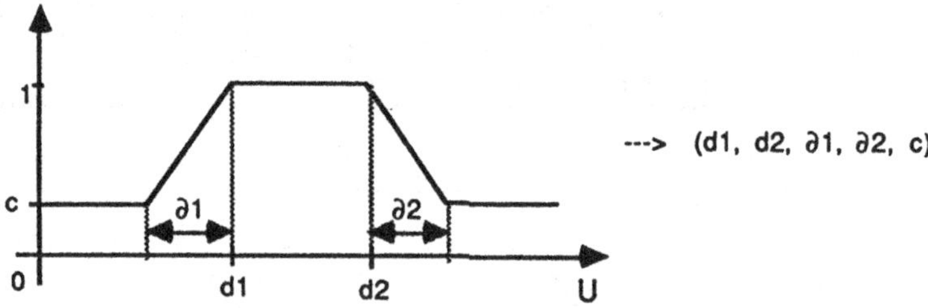

The fact is certain if and only if c=0; the completely possible points are those of [d1,d2], and the points of minimal possibility (c) are U-[d1-∂1,d2-∂2]. Such possibility distributions are sufficient in most of the usual situations; morover, this shape will be preserved in the inference process.

2-the rules

Figure 4 gives half-formalized examples of fuzzy rules processed by TOULMED.

Elementary patterns of the condition parts, as conclusion parts, are restrictions on the value of a variable ("weight-loss rather>= 2 kg", "age rather< 30 years" and also "presumption evidence of NIDD"); so they can be represented by possibility distributions, like facts:

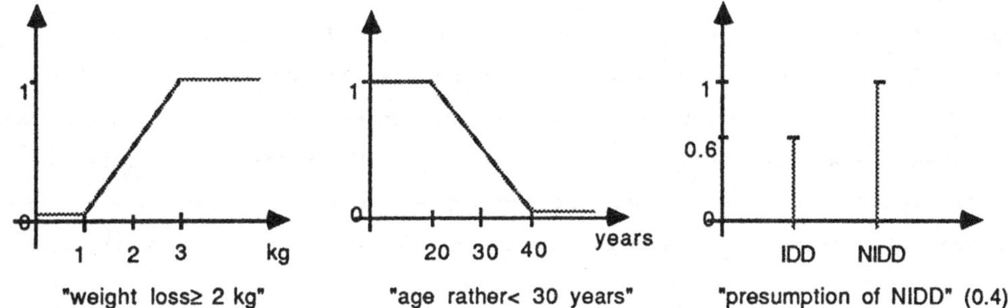

Elementary patterns of condition parts of rules may be imprecise or fuzzy, but generally are not uncertain, since they represent a typical and often ideal situation. On the contrary, the conclusion part may be imprecise or fuzzy, and possibly uncertain; for instance "presumptive evidence of NIDD". In all the cases, possibility theory provides a formalism able to represent such rules. If several elementary pattern are linked by the AND connector (or the OR connector), and if these conditions are *logically independent*, the compound pattern is easily processed (<5>, <8>); in case of dependency, one have to deal conjointly with the dependent variables. It is also possible to take into account the relative importance of the different elementary patterns, by attaching some weights to them, or by modifying the shape of the associated possibility distributions <8>.

Three types of rules have been identified in the collected medical knowledge.

- TYPE 1: rules with non fuzzy conditions (possibly imprecise) and with any kind of conclusion (possibly fuzzy and/or uncertain). e.g. :
sexe = female and age >= 25 ---> ideal weight approximatly= f(height)
In this example, the conclusion is fuzzy but certain.

- TYPE 2: rules with imprecise or fuzzy conditions, and with an uncertain non-fuzzy conclusion. e.g. :
weight-loss rather>= 2 kg and age rather< 30
            ---> presumptive evidence of IDD (0.7)
Here, the conditions are fuzzy, and the conclusion is precise but uncertain.

- <u>TYPE 3</u>: rules with imprecise or fuzzy conditions, and with a certain conclusion, possibly imprecise or fuzzy. e.g. :

    relative-weight rather>= 120% and age rather< 70

            ---> daily-calorie-needs approximately= 1800 cal

All the conditions and the conclusion are fuzzy, but certain.

These three types of rules are allowed in TOULMED. Each of them is processed in a particular way, but within the same theoretical framework. Note that rules with conclusions which are both uncertain and imprecise are not considered, since in practice we can choose to have either a purely fuzzy or a purely uncertain conclusion (due to the balance between imprecision and uncertainty). Indeed an uncertain statement can always be turned into a certain, sufficiently imprecise, statement.

3- application of rules, combination of results

a-propagation of the imprecision and the uncertainty

Inference engines which use crisp rules usually employ modus-ponens as the only inference mode: having a rule A --> B and a fact A, one can detach the conclusion B. An enlarged form of modus-ponens must be used with fuzzy rules: having a fuzzy rule A --> B and a fuzzy fact A', one can detach a conclusion B' which is as close to B as A' is *compatible* with A. If necessary, the contrapositive rule ¬B --> ¬A can be introduced in order to allow modus tollens -like inferences; strictly speaking, the rules A-->B and ¬B-->¬A are no longer perfectly equivalent when they are fuzzy <8>; it is why it is better to elicitate both of them.

Possibility theory provides a *generalized modus-ponens* , which enable us to compute a fuzzy conclusion B', given a rule A --> B and a fact A'. Without explaining all the mathematical details (see <8>, <9>), we can say that this formula is closely linked with a computing tool called *fuzzy pattern matching* <7>, which enable us to estimate to what extent a given possibility destribution D (corresponding to a fact) is compatible with a pattern possibility distribution P (an elementary pattern of a condition of a rule). The fuzzy pattern-matching procedure returns 2 numerical values $\Pi(P|D)$ and $N(P|D)$ belonging to $[0,1]$; $\Pi(P|D)$ estimates to what extent the data D is possibly compatible with what is required by the pattern P, and $N(P|D)$ estimates the certainty of this compatibility, that is to say the degree of impossibility that D is compatible with the opposite requirement. So, in the following case (a):

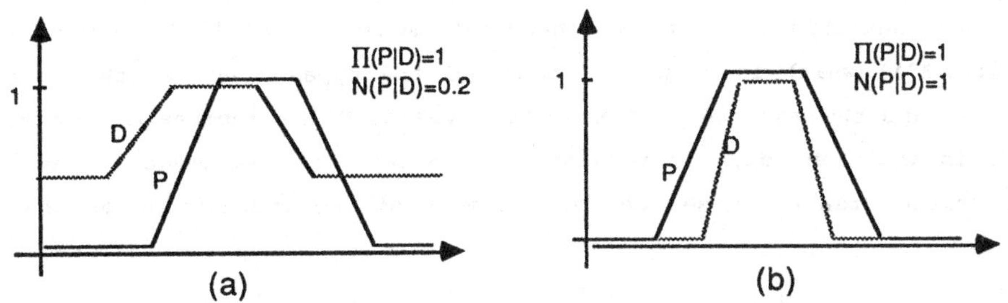

, the compatibility of D with P is completely possible (several values of the domain are completely possible with both P and D), but only somewhat certain. In case (b) on the contrary, the data D is perfectly compatible with the pattern P.

In practice, the generalized modus-ponens has been specialized for each of the three types of rules presented in III-2, for computational efficiency. The figure 8 graphically illustrates the effects of their firing, by showing how the rule conclusion is modified in each case.

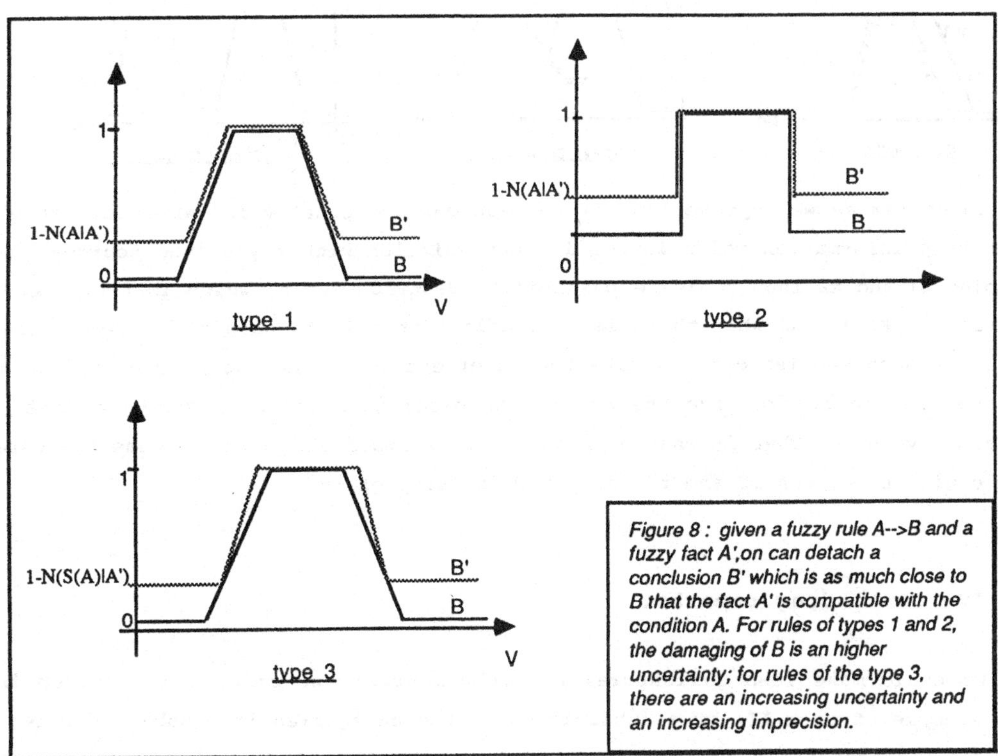

Figure 8 : given a fuzzy rule A-->B and a fuzzy fact A',on can detach a conclusion B' which is as much close to B that the fact A' is compatible with the condition A. For rules of types 1 and 2, the damaging of B is an higher uncertainty; for rules of the type 3, there are an increasing uncertainty and an increasing imprecision.

We can see that this is mainly the necessity measure N(A|A') (more pessimistic than Π(A|A')) which is used. For rules of the types 1 and 2, the smaller this measure, and the more the uncertainty level of B' is increased. For rules of type 3, in addition to an increasing uncertainty, an enlargment of the completely possible area expresses an increasement of the imprecision of the result.

b- combination of information coming from the firing of several rules

Let us suppose that the firing of two distinct rules returns two pieces of information about the same variable Y, in the form of two possibility distributions $\pi 1$ and $\pi 2$. The problem is: how combine the information of those two sources? First we can see if these pieces of information are in conflict. We can measure it by computing the height of the higher point of their intersection. If it is equal to 1, the pieces are not in conflict at all; the close to 0 this value, the stronger the conflict. Those cases are represented in the following figure, where the intersection of $\pi 1$ and $\pi 2$ is pictured by a dotted line.

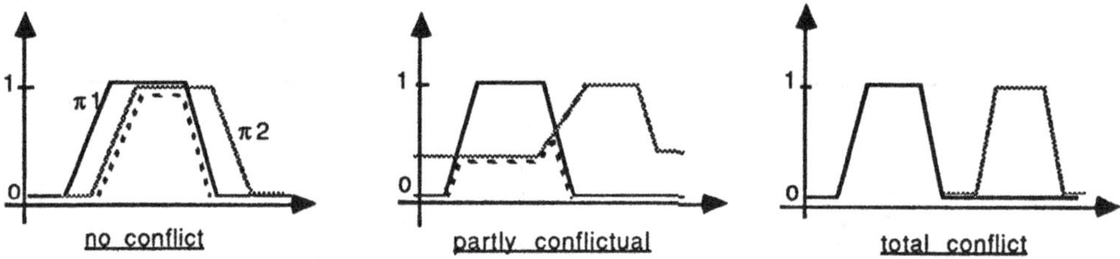

In so far as we suppose that i) the sources are equally reliable and ii) we just keep information which is not in contradiction with any of the sources, we combine $\pi 1$ and $\pi 2$ into a single possibility distribution π, which is the intersection of $\pi 1$ and $\pi 2$ when there is no conflict at all. In case of a slight conflict, a good heuristic is to make the intersection of the two pieces of information, and to *renormalize* the result (in order to still have some completely possible values). That is what happens in the example of figure 5, case 5, where the conflict between r2 and r9 is solved in favor of r9.

IV- Concluding remarks.

We could make less caricatural the behaviour of the crisp rules system in the example of part II, by a subdivision of the categories into sub-categories, and by an increase of the number of rules. But, in addition to the fact that such a cutting is unnatural as regards to the original rules r1, r2,..., a discontinuity of the conclusions will always exist in the vicinity of the boundaries of the sub-categories. Now, if we complete in the example the set of fuzzy rules up to about 15 rules (to deal with more complex medical situations), we obtain a system which make a quasi-perfect diagnosis, according to the experts (but the problem in the example is rather simple, with only some difficult cases).

Fuzzy rules express knowledge in a more synthetic way, by considering a class of situations in a single rule. We can see also that the number of rules used by an expert system is not very significant, without knowing what are their representation capabilities.

Besides, the propagation and the combination of uncertainty and imprecision does not really reduce the speed of the inference engine. The computing time of those computations is indeed much less than the executing time of a cycle of the inference engine; and since a fuzzy rule system often needs less rules than a crisp rule system, it may even run faster on the contrary.

The formulas of possibility theory make essentialy use of the operators *min* and *max*. Consequently, there is no increasing of the errors on the computed possibility degrees when several rules are successively fired (contrastedly with probability theory formulas); moreover it is possible to find out what is the source of the imprecision or the uncertainty of the final results.

REFERENCES

<1>- ADLASSNIG K.P., KOLARZ G., CADIAG-2: "Computer-assisted medical diagnosis using fuzzy subsets, in Approximate Reasoning in Decision Analysis (M.M. Gupta, E. Sanchez, eds), North-Holland, pp. 219-247, 1982.

<2>- BUCHANAN B.G., SHORTLIFFE E.H., "Rule-based Expert Systems. The MYCIN Experiments of the STANFORD Heuristic Programming Project", Addison-Wesley, Reading, 1984.

<3>- BUISSON J.C., FARRENY H., PRADE H., "Un système expert en diabètologie accesible par Minitel", Actes 5èmes Journées Internationales sur les Systèmes-Experts et leurs Applications, pp. 174-189, Avignon, Mai 1985.

<4>- BUISSON J.C., FARRENY H., PRADE H., "The development of a medical expert system and the treatment of imprecision in the framework of possibility theory", Information Sciences, Vol. 37, p 211-226, 1985.

<5>- BUISSON J.C., FARRENY H., PRADE H., "Dealing with imprecision and uncertainty in the expert system DIABETO III" (in French). Proc. 2nd Inter. Conf. on Artificial Intelligence, Marseille, Dec. 1-5, 1986 (CIIAM-86), Hermès Publ., pp. 705-721.

<6>- BUISSON J.C., SOULE-DUPUY C. "Le système INMEDIA d'information pour les diabètiques et leurs médecins", Actes Convention Informatique Latine (CIL-87), Barcelona, Mar. 18-20, 1987, pp. 412-431.

<7>- CAYROL M., FARRENY H., PRADE H., "Fuzzy Pattern Matching", Kybernetes, Vol. 11, pp. 103-116, 1982.

<8>- DUBOIS D., PRADE H. (avec la collaboration de H. Farreny, R. Martin-Clouaire, C. Testemale), "Théorie des Possibilités: Application à la Représentation des Connaissances en Informatique", Masson, Paris, 1985. English version to be published by Plenum (New York).

<9>- DUBOIS D., PRADE H., "The generalised modus ponens under sup-min composition: a theoretical study", in Approximate Reasoning in Expert Systems (M.M. GUPTA, A. KANDEL, W. BANDLER, J.B. KISZKA), North-Holland, pp. 217-232, 1985.

<10>- FIESCHI M., "Intelligence Artificielle en Médecine - Des Systèmes-Experts", Masson, Paris, 1984.

<11>- GASCUEL O., "Un système expert pour la réalisation de diagnostics". Technique et Science Informatiques, 4, 359-372, 1985.

<12>- SOULA G., VIALETTES B., SAN MARCO J.L., "PROTIS, a fuzzy deduction-rule system: application to the treatment of diabetes". Proc. MEDINFO 83, Amsterdam, pp. 533-536.

<13>- ZADEH L.A., "PRUF: A meaning representation language for natural languages", Inter. J. of Man-Machine Studies, Vol. 10, n°4, pp. 395-460, 1978.

COHERENT HANDLING OF UNCERTAINTY VIA LOCALIZED COMPUTATION IN AN EXPERT SYSTEM FOR THERAPEUTIC DECISION.

CARLO BERZUINI

Dipartimento di Informatica e Sistemistica
Universita' di Pavia, via Abbiategrasso 209, 27100 PAVIA (Italy)

GIOVANNI BAROSI, GRAZIA POLINO

Dipartimento di Medicina Interna e Terapia Medica
IRCCS - Policlinico S.Matteo, PAVIA

1. INTRODUCTION

There are problems of therapy selection whose solution crucially depends upon explicit quantification of the uncertainties and tradeoffs involved. In order to handle such uncertainties, some AI researchers have become interested in probabilistically sound methods such as the decision-theoretic approach, rather than relying upon "ad hoc" mechanisms usually available in standard frame/rules-based knowledge representation environments.

Langlotz et.al. [1], for example, use statistical decision theory to evaluate the merit of MYCIN heuristics for therapy selection, and claim that a synthesis of AI and decision theory will enhance the ability of expert systems to provide justifications of their decisions.

Rules in a therapy selection system represent plans for action that may have wide ranging consequences. For example, avoiding the administration of a drug may have the advantage of eliminating undesirable complications, but has the disadvantage of creating the need for another drug, which may have a weaker therapeutic effect and other undesirable effects. The tradeoffs implicit in the choice between alternative therapeutic actions may hardly be broken down into logical steps that unfailingly lead to the best choice, no matter how fine the granularity of the reasoning. In fact the therapeutic expert cannot specify the complete list of features which would guarantee optimal therapeutic choice. Almost always in practice an area of uncertainty is left open.

Relying upon uncoherent mechanisms such as the certainty factor model   for handling such uncertainty is unadvisable.
In fact, the strong independence assumptions entailed by the modularity of such an approach are inherently conflicting with the need of representing tradeoffs.

Among coherent appraches to uncertainty suitable in this context, the one based upon decision trees deserves consideration. Tree analysis leads to recognizing unsual dicision situations, taking appropriate decisions and evaluating n a coherent probabilistic way the relative expected utilities for therapeutic alternatives.

Decision trees, however, may not be the best representation of a problem from the perspective of knowledge modelling, since they do not provide explicit representation of causal dependencies, they are difficult to edit/develop, and are unsuitable to be represented in expert systems.

In section 2, a class of graphical representations of dependencies within a complexly related collection of propositions, the class of **recursive models** , will be described. A recursive model can be represented through a graph whose nodes represent propositional variables, and arrowed links between nodes represent causal influences . This representation provides a visual insight into the qualitative structure of causal processes upon which inference is based.

In section 3, a simple example of how a therapeutic decision problem is modeled via a recursive model is described. The example is drawn from the current development of an expert system for therapeutic advice about anemic patients. The example refers to the decision whether to perform or not splenectomy (= surgical removal of the spleen) in patients affected by myelofibrosis with myeloid metaplasia (MMM). MMM is a chronic myeloproliferative disorder, which may present itself at diagnosis with many possible different scenarios and past stories. Different patterns of disease progression to death may be observed, too. Once a chronic vs. acute characterization of the illness has been established, a complexly structured set of patient symptoms and findings is taken into account for making the decision about splenectomy.

In section 4 the example of recursive model will be compared with a representation based on decision–tree. The greater suitability of the representation by recursive model in the perspective of knowledge representation will be discussed.

The topic of section 5 is propagation of uncertainty on recursive models. Once a quantitative assessment of the strenght of causal influences in the form of conditional probability judgements is available, a simple recursive factorization of the joint probability of the uninstantiated propositional variables conditional on instantiated ones allows propagating uncertainty over the graph, evaluating expected utility and performing various kinds of sensitivity analysis. The computations involved will be illustrated upon the example concerning decision about splenectomy, in section 6.

Section 7 will discuss utility maximization. Section 8 will discuss an approach to propagation of uncertainty by *localized computation,* which makes both the process of probability judgement by the human expert and the machine representation and combination of probabilities particularly feasible. Recursive models are the ideal setting for such an approach.

2. RECURSIVE MODELS

Consider a set of t variables and suppose that the qualitative pattern of causal influences between couples of them is known. These variables form a recursive system if they can be numbered 1 to t , in such a way that

(i)   a certain set of 1 to $k < t$ are nominated as *response variables,*

(ii)   each response variable $i < k$ may depend only on variables $j \in \{i+1,...,t\}$.

Variables $k+1$ to t are said to be *explanatory* and are not caused by any other variable.

We say that a couple of variables i and $j > i$ is unconnected if there is no causal influence of j upon i . In a graph whose nodes correspond to the t variables and where causal influence of variable j upon variable $i < j$ is represented by an arrowed link originating in j and pointing towards i , there will be as many missing links as there are unconnected couples.

The set I^D of missing links and the numbering of the variables entirely specify a recursive model. Within it, the convention is held that each missing link (r,s) with $1 < r < s < t$ and $r < k$ is to be interpreted in the sense that variables corresponding to nodes r and s are *conditionally independent* given all the variables of index $> r$. In the notation proposed by Dawid [2] this is written as $r \perp s \mid \{r+1,...,t\}/s$, or, in a more compact notation, $ZPD(r,s)$, where ZPD stands for Zero Partial Dependence.

For the interpretation of a recursive model the following result is useful.

Consider for each response variable $i \in \{1,...,k\}$ the subset of the variables $i+1$ to t on which it actually depends; this can be divided into a subset $A_i = \{j \mid j > i$ and $(i,j) \in I^D\}$, and a complementary subset of variables $B_i = \{j \mid j > i$ and $(i,j) \in I^D\}$. Then:

PROPOSITION 1. *In the recursive model specified by I^D a response variable $i \in \{1,...,k\}$ is conditionally independent of the variables in B_i given the variables in A_i.*

Now consider the example depicted in Figure 1. Response variables are 1 to 3. The corresponding recursive model has $I^D = (1,4),(2,3)$. Let's apply Proposition 1 with $i = 2$. Since $A_2 = 4$ and $B_2 = 3$, 2 is conditionally independent of 3 given 4, that is $2 \perp 3 \mid 4$.

3. A SIMPLE RECURSIVE MODEL FOR THE CHOICE ABOUT SPLENECTOMY IN MMM.

As an example of recursive model we will consider a simplified version of the model for the decision whether to perform splenectomy or not in an MMM patient, shown in Figure 2.

There are 5 explanatory variables. One of them (SPLEN), denoted by a square, is the dichotomic decision variable with values "splenectomy not performed" and "splenectomy performed".

Remaining explanatory variables, denoted by shaded circles, describe the patient prior to decision: risk class (RISK) with values "low", "intermediate" and "high", is determined using a risk score built over a set of prognostically important variables (see [3]). Pre-operative conditions (PRE-OP) has values "bad" "ok" "very good". Time elapsed since diagnosis (DELTA-T) has values "> 1 year" and "< 1 year". Total erythroid iron turn-over (TEIT), a measure of total erythropoiesis has values "increased", "normal" ,"decreased".

There are 6 response variables: one represents major-disturbances (MD) that can occur as a consequence of the clinical evolution of MMM; values of this variable (md+, md-) respectively denote presence/absence of at least one major disturbance. Variable POST-OP represents possible intra-operative complications, as well as possible complications arising as a direct consequence of splenectomy. Let's assume for POST-OP a simple dichotomy as we have done for MD. Variable RESPONSE has values "positive" and "null": in case of no-splenectomy it is taken to be "null", in case of splenectomy it indicates whether the patient shows positive response to splenectomy, as far as it appears from a complex of haematological signs (eg. decrease of transfusional need). SURV-OP represents survival after operation. SURV>12 and SURV>36 represent survival after 12 and 36 months respectively.

The structure of the model is here presented only for purposes of illustrating recursive models, and is not intended as a definitive result about clinical evolution of MMM.

According to Fig. 2 post-operative complications are more likely to occur, as it is obvious, if pre-operative patient conditions are bad. Positive response to splenectomy is directly connected with expected survival, according to the hypothesis that, for a given risk class a patient with positive response to splenectomy is more likely to survive after 12 or 36 months than a patient with negative response or no splenectomy. Response to splenectomy is made to depend on the time elapsed since diagnosis of MMM (the shorter the better) and on the value of TEIT (the higher the better). Variable MD is

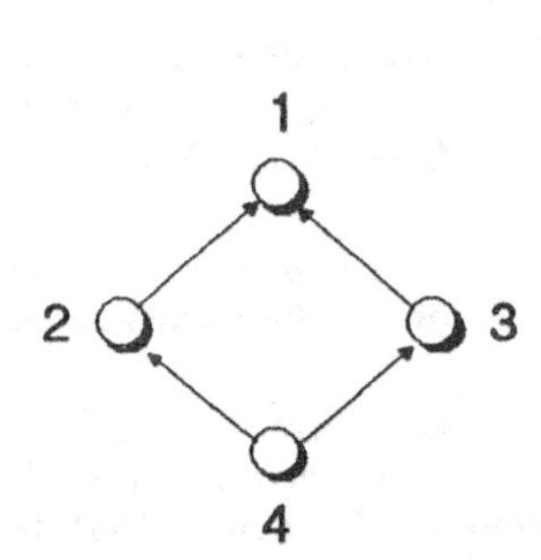

FIGURE 1. Example of recursive model.

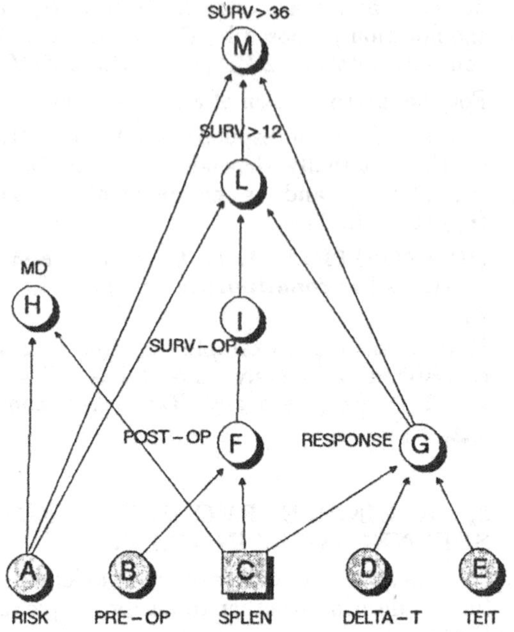

FIGURE 2. RECURSIVE MODEL FOR THE DECISION WHETHER TO PERFORM SPLENECTOMY OR NOT. SHADED NODES DENOTE EX — PLANATORY VARIABLES. SQUARED NODE DENOTES DECISION NODE.

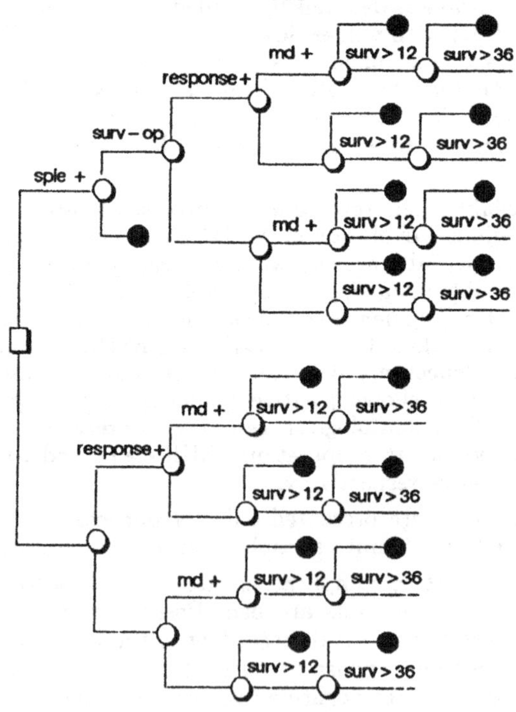

FIGURE 3. DECISION — TREE FOR CHOOSING WHETHER TO PER — FORM SPLENECTOMY.

related to the quality of life. Short and long–term incidence of major disturbances is correlated with patient's risk class and, within a given risk class, may be affected by performing or not splenectomy. Possibly, there is interaction here, in the sense that the strength of the effect of splenectomy on MD may depend upon risk class.

The basic trade–off underlying this decision problem arises because splenectomy may significantly increase expected survival at the cost of a post–operative risk with the concomitant element consisting of an uncertain effect of splenectomy upon quality of life.

ZPD(SURV>36,MD) and ZPD(SURV>12,MD) indicate that risk class is sufficiently predictive of probable future major disturbances that one may predict survival well enough before observing whether these disturbances arise.

Nodes of the graph in Figure 2 have been labeled using upper–case letters. F, G separate future states M, L from past states B, C, D, E. In other terms, once F, G are supposed known, B, C, D, E provide no useful additional information for predicting probability of L, M. In probabilistic terms, this is the Markov property: ZPD(M,B...E) and ZPD(L,B...E).

Certain dependencies loose sense for certain instantiation values. For example, if splenectomy is not performed, G becomes independent of D and E.

4. USING A DECISION TREE FORMALISM

Decision trees provide a formalism to represent decision problems, as well as a coherent mechanism for dealing with the uncertainty associated to these problems. Let's now compare decision trees with recursive models as representation tools.

Figure 3 shows the decision tree for the problem already described via recursive model in the previous section.

Each path through the tree represents one possible combination of therapeutic action, consequences of this action and events that might occur. While the root node represents the decision made by the physician, all other nodes in the tree of Figure 3 represent branch point over which the the physician has no control.

These are indicated by circles and are called "random" nodes.

At every random node the branches starting there have associated with them probabilities of the events corresponding to them, conditional on all the events contained in the tree between that node and the root.

An event is generally represented by different branches in distant parts of the tree. For example, MD+, i.e. occurrence of a major disturbance, is represented by four branches depending on which decision has been made and on RESPONSE. By specifying different values for probabilities attached to these branches, one may model any pattern of conditional dependence among SPLEN, MD, RESPONSE. For example, if the two MD+ branches descending from SPLEN+ have the same probability coefficient, independence of MD on RESPONSE conditional on splenectomy when splenectomy is performed is implied. This gives an idea of how "expressive" a decision tree can be.

However, as revealed by a visual comparison of Figs. 2 and 3, decision trees and recursive models significantly differ as to the capability of *explicit* representation of the pattern of dependencies among events. For example, one cannot deduce from the structure of the tree whether MD+ is influenced by the decision about splenectomy or not, unless he compares numerical values for probabilistic coefficients scattered over the tree. Conditional independencies are even harder to be deciphered.

Difficulty of deciphering causal structure in the tree would have been further increased if the effect of patient's findings such as risk class, TEIT, evident from the recursive model in Fig. 2, would have been modeled in the tree as well. In fact this would multiply the number of branchings. As a consequence, the structure of dependencies among events would be even deeper buried into the tree branching structure.

5. PROPAGATION OF PROBABILITIES ON A RECURSIVE MODEL.

Given a recursive model, a particular form of uncertainty propagation occurs when all and only its explanatory variables are the observable ones. In other terms, all and only the nodes to which no arrows point are instantiated. In this particular situation, propagation of uncertainty is simple. Application of the factorization rule for densities gives:

$$p(1,2,...,k \mid k+1,...,t) = \prod_{i=1}^{k} p_i(i \mid i+1,...,t) \tag{1}$$

which decomposes the joint distribution of the k response variables conditional upon the $t-k$ instantiated explanatory variables into k *component conditional distributions* (CCD). Each of these CCD's may be susceptible of simplification by taking into account the ZPD's. In fact, given $(i,j) \in I^D$, $p(i \mid i+1,...,j,...,t)$ can be simplified to $p(i \mid i+1,...,j-1,j+1,...,t)$. In such a way, the recursive model is decomposed into manageable self–contained subsystems.

In general, by computing $p_k(...)$ to $p_1(...)$, uncertainty is propagated.

In the perspective of knowledge representation, this implies that probabilistic judgements to be elicited from the human expert or obtained from statistical data analysis, that have to be stored in the computer, concern a small number of propositional variables. In other terms, by "local" probabilistic judgements, we achieve "global" coherent propagation of uncertainty.

6. EXAMPLE

With reference to the model represented by Fig. 2, factorization (1) and simplification of the CCD's give:

$$p(MLIHGF \mid ABCDE) =$$

$$p(M \mid LAG) \times p(L \mid AIG) \times p(I \mid F) \times p(H \mid AC) \times p(G \mid CDE) \times p(F \mid BC)$$

A series of conditional probability assessments are now required. For example, 12 numbers are necessary to completely specify $p(M \mid LAG)$, that is the probability of surviving after 36 months for all combinations of values for L A and G. Six of them are zero since "no survival after 12 months" implies zero probability of "survival after 36 months" irrespective of risk class and response to therapy. Let's adopt the convention that lower–case letters represent instantiation values so that, for example, $A = a_1$ denotes "low–risk patient", $L = l$ denotes "survived after 12 months" and $G = g$ denotes "positive response". Suppose that we have:

$$p(m \mid la_1 g) = 0.9$$

$$p(m \mid la\hat{g}) = 0.7$$

This would indicate "high probability of survival after 36 months for a low–risk patient, but higher if a positive response to splenectomy occurs".

We must emphasize that a full assessment of $p(M \mid LAG)$ is necessary, and that this may **not** be derived from separate assessments of, say, $p(M \mid L)$, $p(M \mid A)$ and $p(M \mid G)$. This would lead to the same inconsistencies to which leads the false assumption of independence implicit in the certainty factor model.

Clearly, with respect to a standard certainty factor approach there is a much higher number of probabilistic parameters to be assessed and stored in the computer. One positive aspect of this heavvy parametrization is that this approach compels the human expert to specify probabilities for *all* possible combinations of events. For example, one is forced to specify not only the probability of major disturbances when splenectomy is performed, but also the probability of major disturbances when splenectomy is *not* performed, which does not come naturally with the certainty factors approach in a standard rule–based environment. In fact one might want to specify the rule "IF splenectomy is performed THEN disturbance X is likely to occur (0.4)" *without specifying what happens when splenectomy is not* performed.

Probability is propagated on the graph by first computing $p(f \mid bc)$, $p(g \mid cde)$ and $p(h \mid ac)$. Obviously $p(\bar{f} \mid bc) = 1 - p(f \mid bc)$, a.s.o. Then the *a posteriori* probability for node $p_{POST}(i) = p(i \mid f) \times p(f) + p(i \mid \bar{f}) \times p(\bar{f})$ is calculated. This process continues until the probability for each value of each variable in the model is calculated.

7. MAKING THE DECISION BY MAXIMIZING THE EXPECTED UTILITY

In order to make the decision about splenectomy, both the probabilities of future events influenced by the decision and the desirability of these events should be explicitly considered. Critical events in our example are post–operative survival, survival after 12 months and after 36 months, and quality of life, that is probability of major disturbances. All these events are directly or indirectly influenced by the decision whether to perform splenectomy or not. More in general, there is a subset consisting of n nodes of the graph associated to events on whose probability and desirability the decision has to be based. Let's call these n nodes *target nodes*.

Let X_{ji} denote j^{th} value for i^{th} target node. Suppose that a *utility* $U(X_{ji})$ $i = 1,...,n$, $j = 1,...,m(i)$, where $m(i)$ is the number of possible values of node i, has been determined. Then, for a given set of patient's data, the decision is made by maximizing expected utility, which is expressed as follows:

$$Expected\ \ Utility\ =\ \sum_{i=1}^{n}\sum_{j=1}^{m(i)} p(X_{ji}) \times U(X_{ji})$$

8. PROPAGATION BY LOCALIZED COMPUTATION

Recursive models provide an ideal setting for a particularly attractive process of combination of probabilities for propagating uncertainty by *localized computation*. Recursive models naturally lend themselves to implementation by frames: each node of the model, in fact, might be represented by a frame. In a frame for node i, slots provide a place for storing:

(a) the set of probabilistic parameters which specify the CCD $p(i \mid ....)$

(b) current probabilities for all possible values of variable i;

(c) "pointers" to predecessor frames, i.e. those corresponding to predecessor nodes in the causal recursive model;

In order to propagate uncertainty it is sufficient that each frame i inspects predecessor ones to know the probability distribution for values of predecessor variables and activates the computation of probabilities of values of variable i.

This process may occur by "message sending" between frames. Sending a "message" to frame i returns probabilities for values of variable i. A frame to which the message is sent may have to send messages to predecessor ones for necessary information, thus triggering a recursive process that ripples from the frame to explanatory nodes and back in order to return requested probabilities.

The important characteristics of such a computational scheme for probability propagation are:

(1) all probabilistic information concerning a variable is "localized" since it is stored in a single frame representing that variable;

(2) propagation occurs by message sending between frames

"Localization" of probabilistic information has many advantages. For example, simple accessing of a frame allows the user to change probabilistic parameters he wants to change; then, by sending a message to a frame he sets up a process of updating the probability distribution of the corresponding variable (and possibly of the overal utility). As a generalization of this concept, sensitivity analysis can easily be performed. Such a kind of interaction is enhanced by the fact that the recursive model lends itself to representation by images on the screen, so that the user can reason on it in order to decide which parameters to change.

Nodes of the recursive model, modeled as frames, may be part of hierarchies so that for example, with reference to Fig. 2, node MD can be further refined into a hierarchy of disturbances.

9. REFERENCES

[1] Langlotz, C.P., Shortliffe, E.H., Fagan, L.M., Using Decision Theory to Justify Heuristics, Proceedings of AAAI–86, Fifth National Conference on Artificial Intelligence, Philadelphia, PA, August 1986, pp. 215–219.

[2] Dawid, A.P., Conditional Independence in Statistical Theory (with discussion), J.R.Statist.Soc.B, vol.41,pp.1–31,1979.

[3] Barosi, G., Berzuini, C., Palestra, P., Polino, G., A Classification of Myelofibrosis with Myeloid Metaplasia Based on Prognostic Factors, (submitted for publication).

MUNIN - On the Case for Probabilities in Medical Expert Systems - a Practical Exercise.

--o--

Finn V. Jensen[1], Stig K. Andersen[2,3],
Uffe Kjærulff[1], and Steen Andreassen[2,3]

[1]Judex datasystemer, Lyngvej 8, DK-9000 Aalborg, Denmark
[2] Nordjysk Udviklingscenter, Badehusvej 23, DK-9000 Aalborg, Denmark
[3]Institute of Electronic Systems, Aalborg University,
Strandvejen 19, DK-9000 Aalborg Denmark

Abstract.

MUNIN - an expert system for electromyography - provides a practical demonstration that an expert system of non-trivial size based on Bayesian probability theory can be constructed. Much of the medical knowledge in the system is embedded in a causal probabilistic network. The nodes in the network represent medical concepts, diseases, pathophysiological states or findings, and the links between the nodes govern the interaction. The inference method is an adaption of the method developed by Kim and Pearl (1983). The system provides the user with a set of spreadsheet-like tools to assist him in the decision process. The user is allowed to explore the effect of adding, retracting or changing findings. The time required to update all probabilities in the network is about 10 seconds in the current implementation†. Specialized tools are provided to allow the user to inspect where evidence for a node comes from. This tool also allows the user to trace the origin of conflicting evidence. To assist the user in the planning of test sequences, an importance mechanism was provided. This mechanism assigns importance to individual findings, which reflects their capacity to provide information about other nodes in the network.

---

*This work is supported in part by the EEC ESPRIT programme, project P599
† Interlisp on a XEROX 1109.

1. Introduction

The current scarcity of expert systems where the reasoning is based on Bayesian probability theory may be due to misconceptions about probabilities found in the literature. As argued by Cheeseman (1985), these misconceptions have led to the attitude: "The Bayesian approach doesn't work - so here is a new scheme". Several of these expert systems based on ad hoc "probability" concepts have been successful in a number of ways, demonstrating the necessity of being able to handle uncertainty in medical expert systems. They also demonstrate the need for a theoretically sound handling of uncertainty.

In Andersen et al. (1986) it was postulated that knowledge organized in a causal network can be used for a unified approach to the main tasks of a medical expert system: diagnosis, planning of tests and explanations. The present paper explores this postulate in a causal probabilistic network. It also provides a practical demonstration that the problems supposedly associated with probabilistic networks are either non-existent or that practical solutions can be found.

This paper reports on the methods implemented in MUNIN* -an expert system for electromyography (EMG) (Andreassen et al. 1987). EMG is the diagnosis of muscle and nerve diseases through analysis of bioelectrical signals from muscle and nerve tissue. In Andreassen et al. (1987) the MUNIN prototype network was presented, and the problems associated with knowledge acquisition, knowledge representation and knowledge verification were discussed. In this paper the methods for propagation of evidence, based on the work of Kim and Pearl (1983) will be outlined, and issues like planning of test sequences, explanation of the line of reasoning, tracing of conflicting evidence and user interfacing will be discussed. The proposed user interface provides the user with a spreadsheet-like graphical interface to the system.

2. The domain

In the MUNIN prototype, much of the medical knowledge is represented in a causal probabilistic network. A causal probabilistic network consists of a set of **nodes** and a set of directed causal **links**.

The domain is roughly divided into three levels, a disease level, a pathophysiological feature level, and a findings level. The three levels are linked by causal relations: Diseases cause certain affections in muscles. These affections in turn cause expectations for certain findings. Intermediate nodes between the levels can occur.

---

*MUNIN: MUscle and Nerve Inference Network. According to Norse mythology, Munin is one of the two ravens, whispering intelligence into the ear of the god, Odin.

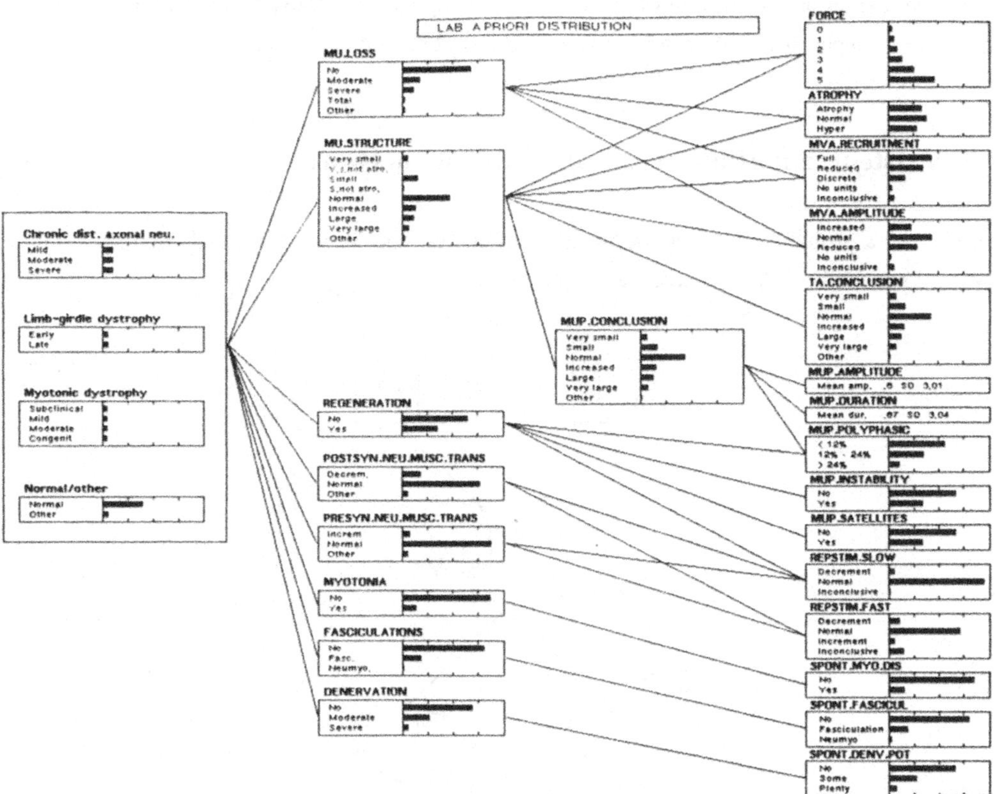

Fig. 1. The nodes and links in the MUNIN prototype. The a priori probabilities for the individual disease states (left) are propagated through the layer of pathophysiological nodes (middle) to the finding nodes (right). The length of the horizontal bars indicates the probability of the states in the nodes.

In the MUNIN prototype the disease level has degenerated into a single node, the D-node. This node has 11 states (see fig. 1), representing different grades of three different diseases, plus the states normal (no neuro-muscular diseases) and other (a disease different from the three diseases). The disease node has causal links to 8 pathophysiological nodes (P-nodes) describing the pathophysiological status of a muscle. The P-nodes are linked either directly or through intermediate states to 15 nodes representing findings or measurements (F-nodes).

Most of the nodes have a discrete set of states, with 2 to 9 states. However, the F-nodes MUP.AMPLITUDE and MUP.DURATION have a continuous state space.

We believe that an understanding of the medical concepts in the domain is not necessary for the reading of this paper. If not, the reader is referred to Andreassen et al.(1987).

3. Consistent updating of probabilities in networks.

Each node in a probabilistic network has a set of **states**, e.g. V: $v_1, \ldots v_v$, and each state has an associated probability, $P(v_i)$, where $\Sigma_{i=1,v}P(v_i) = 1$.

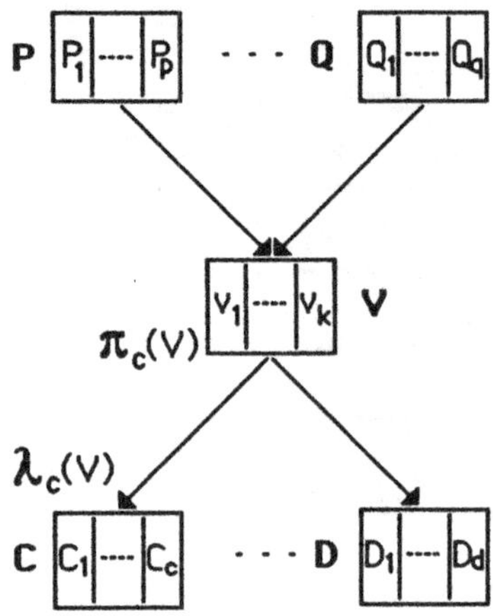

fig. 2. A typical section of a causal network.

The causal links between nodes are expressed as conditional probabilities. In fig. 2, the links impinging on V have the associated conditional probability of V being in state v_i, given that the nodes P, . . ., Q are in the states $p_j, \ldots, q_k$: $P(v_i| p_j, \ldots, q_k)$. The nodes P,..,Q in fig. 2 are called parents of V, and C,..,D are called children of V.

Practical ways of obtaining these conditional probabilities, utilizing deep knowledge of the domain to construct small probabilistic models, are outlined in Andreassen et al. (1987).

Currently there is no computationally efficient way of updating probabilities in a causal probabilistic network with arbitrary topology. However, if the network is **undirectionally acyclic** and if **conditional independence** is assumed, i.e. $P(c_j, \ldots, d_k \mid v_i) = P(cj \mid v_i) \ldots \ldots P(d_k \mid v_i)$, then Kim and Pearl (1983) has proposed an efficient method for updating of probabilities. The idea behind their method is that any link in an acyclic network divides the network into two parts. The link is the only channel for flow of information between the two parts. What they achieved was to define concepts, **causal evidence** π and **diagnostic evidence** λ, in such a way that the probabilities of the states of a node can be calculated as a normalized product of π and λ.

For example, the probability distribution for the states of V, calculated only from evidence coming from the part of the network containing the child C and affecting V through the link (V, C) is by Kim and Pearl called the diagnostic evidence $\lambda_C(V)$. The probability distribution over the states of V, calculated from the evidence in the rest of the network, excluding the part containing C, they call the causal evidence $\pi_C(V)$.

3.1 Adapting the MUNIN network to the constraints.

The MUNIN prototype network does not meet the constraint of being acyclic. In fig 3. two different types of cycles are illustrated. Both types of cycles can be found in the MUNIN network.

Methods for propagation of probabilities in networks with cycles are currently being developed (Pearl, 1986; Lauritzen and Spiegelhalter, 1987). Meanwhile, we have chosen to modify the network in order to meet

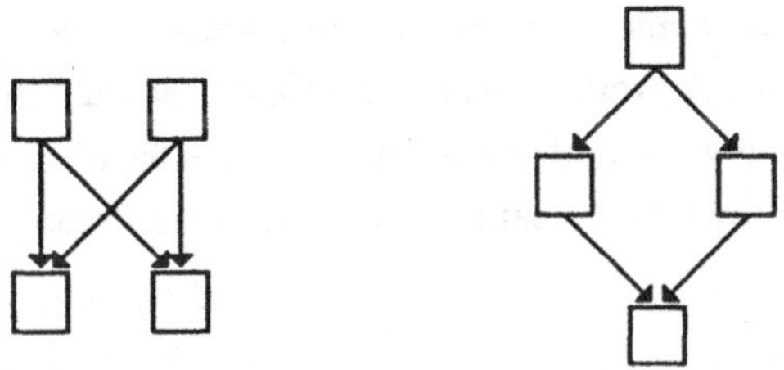

Fig 3 Examples of cycles in a network

the constraints. For example, the cycle MU.STRUCTURE-ATROPHY-MU.LOSS-FORCE -MU.STRUCTURE was eliminated by combining the two nodes MU.LOSS and MU.STRUC-TURE with 5 and 9 states respectively into a supernode MU.LOSSxMU.STRUCTURE with 5x9 states. The probability distributions of MU.LOSS and MU.STRUCTURE can then be calculated as the marginal distributions of MU.LOSSxMU.STRUCTURE

The nodes REGENERATION, POSTSYN.NEU.MUSC.TRANS and PRESYN.NEU.MUSC. TRANS were also combined into a supernode. The remaining cycle DISEASE-RE-GENERATION-MUP.POLYPHASIC-MUP.CONCLUSION-MU.STRUCTURE-DISEASE was eliminated by removing the link from MUP.CONCLUSION to MUP.POLYPHASIC. This link was a "weak" link, and the removal of the link only represents a minor distortion of the medical knowledge.

3.2 Initialization of the network.

To initialize the network, dummy nodes were added. The disease node was given a dummy parent node, D_d with the same states as the disease node and the findings nodes were given a dummy child node (F_d) with the same states as the parent. During the initialization of the network, the a priori probability distribution of diseases was assigned to the D_d node and even distributions were assigned to the F_d nodes. This is consistent with the Maximum Entropy approach underlying the probability propagation algorithm developed by Kim and Pearl (1983). It has been proved (Jensen et al. 1987) that if and only if probability distributions are assigned to all the dummy nodes, a consistent (and unique) set of probability distributions can be calculated for all nodes in the network. Therefore the distributions calculated by the system are independent of the order in which information is entered, and from the system's point of view no information is inconsistent.

4. User Interface

During the course of an EMG examination, findings are made available to the network. Each time a finding is made available, the network updates the BELIEF probability distributions for all the nodes in the network. Hopefully, the probability distribution in the disease node gradually changes from the a priori distribution towards a situation where one of the disease states in the disease node

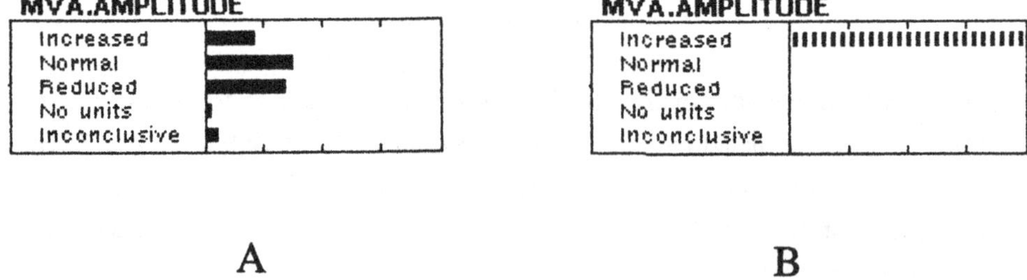

Fig 4 A: a findings node before the finding has been entered. The current BELIEF distribution for the states is indicated by the length of the bars. B: a findings node after the finding has been entered. The outcome of the measurement is indicated by the dotted bar at the actual state.

has such a high probability that the examiner feels that he has arrived at a diagnosis. If this goal is not achieved, then the EMG examination is finished when none of the remaining EMG tests have any significant chance of establishing a diagnosis.

When a finding is made available to the network, this is signalled graphically to the user. Before a finding is entered into the network, the findings node displays the expectations for that finding (fig. 4). The horizontal bars indicate the BELIEF probability distribution for the states in the node. When a finding is entered, a broken 100% bar indicates the outcome of the test.

Findings can be made available to the network through two different mechanisms. A computer based EMG measurement system is an integrated part of the system. During the course of an EMG examination, the EMG measurement system will transmit measurements to the network through a 'Data Capturing System',which is an intelligent interface to the network. The Data Capturing System has access to normal values which it uses to convert measurements into findings. This automatic data entry system was devised to reduce the time the user has to spend on communication with the expert system, a consideration that we think will ultimately determine whether or not the system is acceptable to the user.

Besides this automatic system, the user may manually enter findings directly into the findings nodes. He may also manually retract or change findings. This provides the user with a spreadsheet-like facility which he may use to study the effect of adding or deleting a finding, for example to explore if a finding can affect the diagnosis.

5. Tracing of evidence

Even though the above mentioned facility for manual entering of findings opens up the possibility for exploring the network without performing the actual EMG tests, the user may wish a tool that gives him direct access to the evidence that contributed to the final BELIEF in the states of a node. The function of such a tool is explained through an example.

If the user wants to assess the evidence that gave rise to the current BELIEF in the states of the node MUP.CONCLUSION (abbreviated CON), then a diagram showing the adjacent nodes of CON will be shown (fig 5.A). The probability bars inside the box MUP.DURATION (abbreviated DUR) do **not** show the distribution of probabilities for the states of DUR. They show the probability distribution for CON given only the evidence coming through DUR. Recalling the definition of diagnostic evidence, this is identical to $\lambda_{DUR}(CON)$. Similarly, the box MUP.AMPLITUDE (abbreviated AMP) displays $\lambda_{AMP}(CON)$. The box MU.STRUCTURE (abbreviated STRUCT) in a similar way displays the evidence for CON coming through STRUCT ($\gamma_{STRUCT}(CON)$).

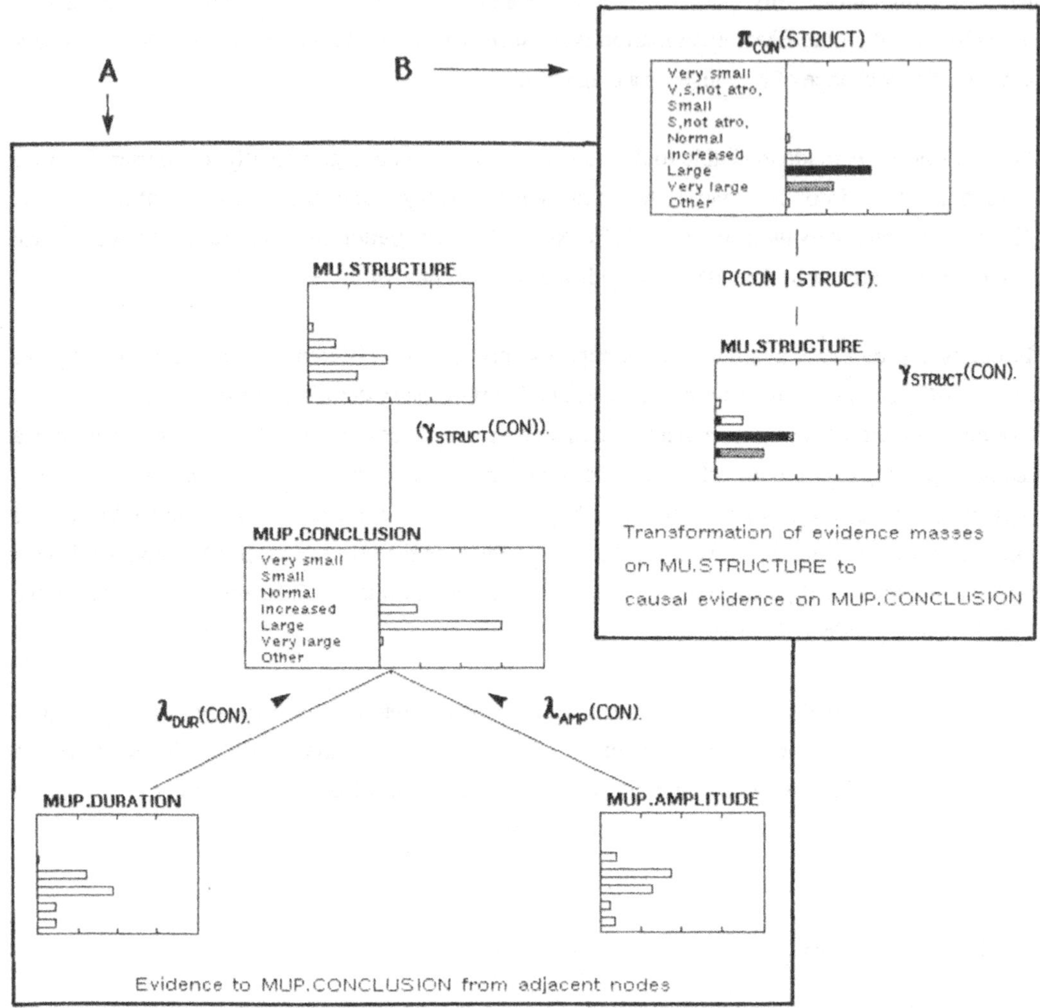

fig 5   A: Evidence for MUP.CONCLUSION. B: The transformation of evidence for MU.STRUCTURE to evidence for MUP.CONCLUSION

The user might at his next step want to know how the evidence from the adjacent nodes was derived. He can get that. Fig 5.B is shows how the causal evidence for STRUCT is transformed to $\gamma_{STRUCT}(CON)$. It is obtained by multiplying $\pi_{CON}(STRUCT)$ with the conditional probability matrix P(CON | STRUCT). The shadings in fig 5.B are used to keep track of individual terms in the product sum - or, in other words, to illustrate which states in CON are supported by which states in STRUCT.

The situation is a little more complex when either the node being considered or its children has more then one parent. The evidence from one parent can not be evaluated independently of the evidence from the other parents. By making reasonable assumptions about other parents, the method outlined above can be generalized to a situation with multiple parents (Jensen et al. 1987).

6 Importance

To help the EMG examiner to decide which EMG test to perform next, a planning tool is implemented.

We have addressed the problem in the following way: Given that some findings have been entered already, and given that the user is interested in a particular node V. Which F-node is most **important** in the sense that it will probably provide the most information on V ?

In information theory, the concept of entropy is a measure for amount of information. The entropy of a node V is defined by

$$H(V) := \sum_i -BELIEF(v_i) \, \log_2(BELIEF(v_i))$$

$H(V)$ is the amount of information which V can receive before the state of V is completely determined. Let N be a finding node with states $n_1,..,n_k$. Let $BEL(v_i \mid n_j)$ denote the BELIEF of the state v_i if the information that N is in state n_j is entered into the network. The average change of entropy for V if node N becomes fully informed, $I(V,N)$, is

$$I(V,N) := \sum_i \sum_j BELIEF(n_j) \, BEL(v_i \mid n_j)[\log_2(\, BEL(v_i \mid n_j)/BELIEF(v_i))]$$

$I(V,N)$ is called the **mutual information** between V and N, and it should be thought of as the amount of information which in average will be propagated to V if the state of N becomes known.

By **importance** of N for V, $IMP(V,N)$ we mean the information which in average can flow from N to V, given as a percentage of information which V can receive totally.

Fig 6. shows the result of a request for importance for the node MU.STRUCTURE. The importance figures in fig 6. suggest that a Motor Unit Potential test should be performed rather than a Turns and Amplitude test. Fig. 7 shows the probabilities in the network after this has been done.

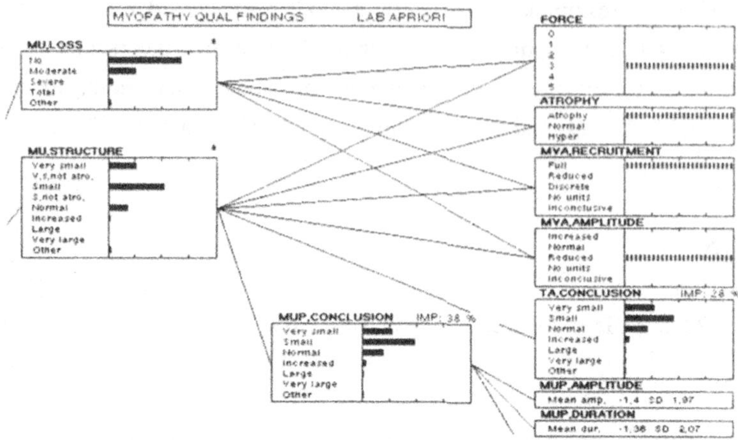

Fig 6. A screen dump of the LOP, when importance for MU.STRUCTURE is requested (indicated by a *) . The importance is shown at the top right of the findings nodes.

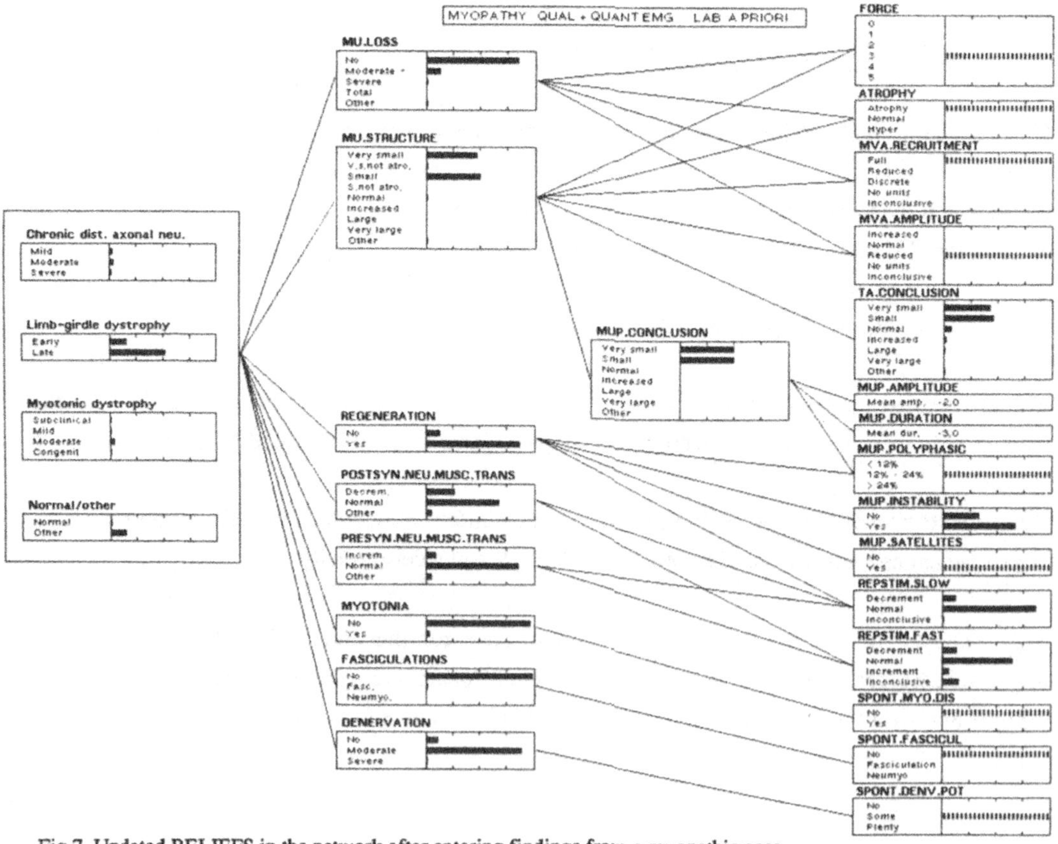

Fig 7. Updated BELIEFS in the network after entering findings from a myopathic case.

6.Conclusion

Though the MUNIN prototype only represents an implementation of a small network, and its current form only has limited functionality, it allows us to reach a number of conclusions.

First of all, we conclude that using a causal network as the core knowledge base for an expert system does not delimit the diversity of system functionality compared to e.g. rule based systems.

Besides, the network representation of knowledge provides a clear distinction between domain knowledge and inference methods. Furthermore, the functionality implemented so far has the advantage of having a sound theoretical base - that of Bayesian probability theory.

Response times in the system are far from being frightening, and we find our experience very encouraging for the further work on treating probabilities in medical expert systems as what they are, namely probabilities.

Acknowledgements:

We thank the co-workers in the MUNIN group. These are B. Falck, A. Rosenfalck, A.R. Sørensen and M. Woldbye. Further, we thank the other participants in the ESPRIT P599 project. They come from the companies Computer Resources International, Copenhagen, and LOGICA, Cambridge. We have had many valuable discussions in the project group, by which many of the ideas behind our implementation have matured.

7. References

Andersen et al. (1986).
   S.K.Andersen, S.Andreassen and M.Woldbye: Knowledge Representation
   for Diagnosis and Test Planning in the Domain of Electromyography,
   Proc. 7th Europ. Conf. on Artificial Intelligence. Brighton, UK, July 1986, pp 357-368.

Andreassen et al. (1987).
   S.Andreassen, M.Woldbye, and B.Falck: MUNIN -A Causal Probabilistic
   network for interpretation of electromyographic findings,
   Proc. 10th International Joint Conference on Artificial Intelligence, Milan ,Italy, August 1987.

Cheeseman (1985)
   P.Cheeseman: In Def Los Angeles, California, August 85, pp 1002-1009.

Kim & Pearl (1983)
   J.H.Kim and J.Pearl: A Computational Model for Causal and Diagnostic
   Reasoning in Inference System in Proc. 8th Joint Int. Conf. on Artificial
   Intelligence, Karlsruhe, West Germany, pp 190-193.

Lauritzen & Spiegelhalter (1987)
   S.L.Lauritzen and D.Spiegelhalter: Fast Manipulation of Probabilities
   with Local Representations - With Applications to Expert Systems.
   Research report R-87-7, Institute of Electronic Systems, Aalborg University, Denmark,
   ISSN 0106-0791.

Jensen et al. (1987)
   F.V.Jensen, S.K.Andersen, U.Kjærulff, S.Andreassen:
   A Causal Network Prototype in the Domain of Electromyography - an
   Implementation of Coherent Probabilistic Methods.
   Research report R-87-8, Institute of Electronic Systems, Aalborg University, Denmark,
   ISSN 0106-0791.

Pearl (1986)
   Judea Pearl: Fusion, Propagation, and Structuring in Belief Networks, in
   Artificial Intelligence 29 (1896) pp 241-288

RULE BASED EXPERT SYSTEMS IN GYNECOLOGY:
STATISTICAL VERSUS HEURISTIC APPROACH

P.A. Riss, H. Koelbl, A. Reinthaller, J. Deutinger

2nd Department of Obstetrics and Gynecology, University Hospital,
Spitalgasse 23, A-1o9o Vienna, Austria

INTRODUCTION

Any area of medicine, in which the number of answers to a particular question - e.g. the number of possible diagnoses - is limited,
and in which it is possible to structure the line of reasoning is
well suited for the use of computer assisted medical decision making
(Reggia and Thurim, 1985).

Before starting with an expert system, however, one has to
decide which strategy to adopt. We have previously reported our experience with rule based expert systems in the field of urodynamic
diagnosis and cycle stimulation (Riss and Koelbl 1986, Riss et al
1987). In the present paper we want to review two possible approaches
- statistical and heuristic - in the construction of rule based
expert systems for limited applications in clinical medicine.

HARDWARE AND SOFTWARE

We used an IBM-PC with 2 disk drives and 256k RAM. The expert
systems were developed using the expert system shell EXSYS (Exsys
Inc, Albuquerque, New Mexico, U.S.A.), a rule based system consisting
of an editor to develop and edit the knowledge base (the rules), and
a RUNTIME program to run the expert system. EXSYS uses backward
chaining to arrive at the correct answer.

The "expert" has to develop the knowledge base, i.e. he has to
define and write the rules. The structure of the rules is
"IF...THEN...", in which the THEN part is either one of the
solutions or a statement which is used in a subsequent rule.

All possible solutions - called "choices" - have to be defined at the outset and must be found in the THEN part of at least one rule. In addition, the expert can assign probabilities to the different choices.

The system asks specific questions which are answered either by giving one of several answers (multiple choice system) or by entering numeric data. Notes and references can be added to every rule to provide additional information on each rule and to explain how the expert system arrived at the solution.

STRATEGIES FOR DEVELOPING THE EXPERT SYSTEM

In building an expert system - and in particular a rule based expert system - one has to follow certain steps (Adlassnig et al 1985). After defining the application the first step consists in identifying the possible outcomes (choices, solutions, diagnoses, suggested actions, etc). The next step is the selection of the variables which will be used in the expert system: the variables should be easily and consistently available, reliable, and actually used in clinical work. Then the rules have to be written, before the expert system can be tested and modified. A very important point which will not be addressed in the present paper, is maintenance and regular update of the expert system.

THE STATISTICAL APPROACH: URODYNAMIC DIAGNOSIS

Urodynamic examination is done preoperatively to make a differential diagnosis of urinary incontinence in incontinent women. As the first step we defined the 5 possible diagnoses: genuine stress incontinence, motor urge incontinence, sensory urge, mixed incontinence, and no incontinence demonstrable.

All urodynamic investigations of the years 1984 and 1985 per- formed at our department were used for the development of the expert system, the only condition being that the patient was incontinent and that the urodynamic examination was complete (N=465). The prevalence of the different diagnoses was as follows:

stress incontinence 5o.5%, motor urge incontinence 1.7%, sensory
urge 2.6%, mixed incontinence 18.3%, and no incontinence demonstrable
26.9%. The variables obtained at urodynamic examination were used
for the construction of the expert system.

In order to derive the necessary rules we calculated the predictive value of each available variable for each of the five urodynamic diagnoses. For every diagnosis we selected the 4 variables
with the highest predictive values and designed a tree diagram,
using the variables in the order of their predictive values (Riss
and Koelbl 1987, Wasson et al 1985). This process of building an
empirical tree diagram by repetitively splitting the patient
population into smaller and smaller categories is called recursive
partitioning analysis (Friedman 1977).

For each diagnosis we thus obtained several subsets of
patients, and for each subset we calculated the predictive values
of the combination of urodynamic variables characterizing the subset.
We used the combinations of variables for each subset to develop a
rule (IF....AND...AND...THEN DIAGNOSIS X PROBABILITY Y) and assigned
to the diagnosis the probability calculated as the predictive value
for the given subset.

THE HEURISTIC APPROACH: STIMULATION OF THE MENSTRUAL CYCLE

In a second application we developed an expert system for
stimulation of the menstrual cycle in an in vitro fertilization
program. The menstrual cycle was stimulated with clomiphene and
human menopausal gonadotropin (HMG), and ovulation was induced with
human chorionic gonadotropin (HCG). During stimulation the patient
is monitored by daily serum hormone level determinations and ultrasound examination. 7 variables were used for the expert system, the
end point was one of 3 possible lines of action: 1) continue stimulation 2) induce ovulation, and 3) cancel treatment cycle. These
solutions are mutually exclusive, we also did not add probabilities
(or certainty factors) to the solutions.

We now wrote a total of 28 rules, trying to cover every
possibility for the cycle days 8 to 14. The rules as well as the

order in which the rules were placed in the knowledge base, i.e. the expert system, were decided by the developers and based entirely on experience and intuition.

COMMENT

We present two possible approaches to the development of small rule based expert systems for clearly defined applications. The statistical approach offers the advantage of yielding probabilities for the different solutions, and these probabilities can be combined. On the other hand, the major drawback of the purely statistical approach is that every combination of variables has to exist in the study group, and that it is impossible to calculate relevant predictive values for diagnoses with a low prevalence in a given sample, since the predictive value depends on the prevalence of a diagnosis (Stempel 1982).

The purely heuristic approach - as used in the expert system for cycle stimulation - does not permit the assignment of calculated probabilities. In addition, it is almost impossible for an expert to think of all possibilities and to cover every aspect of a solution.

Both expert systems presented here are in actual use at our department. In summary, we can draw several conclusions with regard to the development and the use of the expert systems:
1) it is very difficult to cover every aspect of a problem with an expert system. While the statistical approach chosen in the first application assigns certainty factors to every combination of variables seen in the last 2 years, the heuristic approach to cycle stimulation can only be refined through continued testing; even then the developer can never be sure that he has thought of all possibilities.

2) we doubt that an expert system for diagnosis or treatment can be better than a human expert. An expert system might give a more complete list of solutions, or might suggest solutions the physician may not have thought of, or might do calculations which are difficult or time consuming to do by hand. However, the essential aspect of an expert is his (or her) ability to learn from

experience and to evaluate a situation on the basis of a year long
and constantly updated experience. In contrast, most expert systems -
and certainly simple rule-based systems - are not able to learn.

3) on the other hand, expert system are very valuable in
teaching situations: the different steps in the development of a
diagnosis can be clearly shown and explained.

4) last not least, expert systems can be very helpful when a
"true" expert is not readily available.

REFERENCES

Adlassnig, K.P., Kolarz. G., Scheithauer, W., Effenberger, H. &
     Grabner, G. (1985) CADIAG: Approaches to computer-assisted
     medical diagnosis. Comp Biol Med 15, 315-335

Friedman, J.H. (1977) A recursive partitioning decision rule for
     nonparametric classification. IEEE Trans Comput 16, 4o4-4o8

Reggia, J.A. & Thurim, S. (1985) An introduction to computer-
     assisted medical decision making II. MD-Computing 2: 4o-46

Riss, P. & Koelbl, H. (1986) Strategien zur Entwicklung eines
     Expertensystems am Beispiel der urodynamischen Untersuchung.
     In: Medizin-Technik Medizinische Informatik'86. Eds. by
     Rappelsberger, P., Pfundner, P. & Gell, G. p. 351-355.
     R. Oldenbourg, Wien-München

Riss, P. & Koelbl, H. (1987) The predictive values of single and
     combined urodynamic parameters. Acta obstet gynecol scand
     (submitted for publication)

Riss, P., Reinthaller, A. & Deutinger, J. (1987) An application of
     computer-assisted medical decision making: cycle stimulation in
     an in vitro fertilization program. J Perinat Med 15 (Suppl 1):15

Stempel, L.E. (1982) Eenie, meenie, minie, mo... What do the data
     really show? Am J Obstet Gynecol 144: 745-752

Wasson, J.H., Sox, H.C., Neff, R.K. & Goldman, L. (1985) Clinical
     prediction rules. Applications and methodological standards.
     N Engl J Med 313, 793-799

Knowledge Engineering Tools

A Radiological Expert System for the PC - Design and Implementation Issues

W.Horn[*), H.Imhof[+), B.Pfahringer[*), E.Salomonowitz[+)

[*)Department of Medical Cybernetics and Artificial Intelligence,
and
[+)Central Department for Radiological Diagnostics,
University of Vienna Medical School, Austria

Introduction

The development of an expert system for decision support in radiological diagnosis (RADIO) is a joint project of the Department of Medical Cybernetics and Artificial Intelligence and the Department of Radiological Diagnostics of the University of Vienna.

The goal of the project is to design a very user friendly system implemented on the personal computer. At any time during system usage the expert radiologist has to be able to use and to modify the knowledge represented in the system. This results in the need for three things:
- a transparent knowledge representation scheme, that structures knowledge in a way familiar to the radiologist;
- the implementation of a knowledge acquisition and knowledge maintenance component, that supports the radiologist. Oriented towards the design of a comfortable 'knowledge inspection tool' [1], it has to present him all pieces of knowledge in the system, showing all the relations that exist between these pieces of knowledge, and gives him an instrument to add new knowledge and modify the existing pieces of knowledge. In addition, the system has to prevent inconsistencies that may arise from the application of the basic mechanism inherent to the selected representation scheme;
- the development of a consultation module that supports the user in an efficient way entering the symptoms of a specific patient.

The development of the expert system RADIO is a long term project that is planned to run through several cycles of modification, enhancements, knowledge refinement, and knowledge structure changes. Currently we are in the second cycle.

As a consequence, the system has to be implemented using a most flexible tool, both, for knowledge structure enhancements, and for defining new methods manipulating the knowledge. We have chosen VIE-KET (Vienna Knowledge Engineering Tool), a frame-based system developed at the Austrian Research Institute for Artificial Intelligence [2,3]. The decision in favor of a frame-based tool was mainly influenced by experiences with the expert system ESDAT [4,5]. In addition, frames seem to reduce the distance between the internal model of a domain and the cognitive model of the expert [6].

We have started with two domains: diseases of the kidney and the urinary tract, and mamma diagnosis. Currently we are focussing on the kidney. The main emphasis is put on representing the knowledge of expert radiologists gained through many years of practical experience. The elementary content of the knowledge base has been built using a catalogue that defines the requirements for a radiologist examination [7]. This knowledge has been and is extended and modified by the expert.

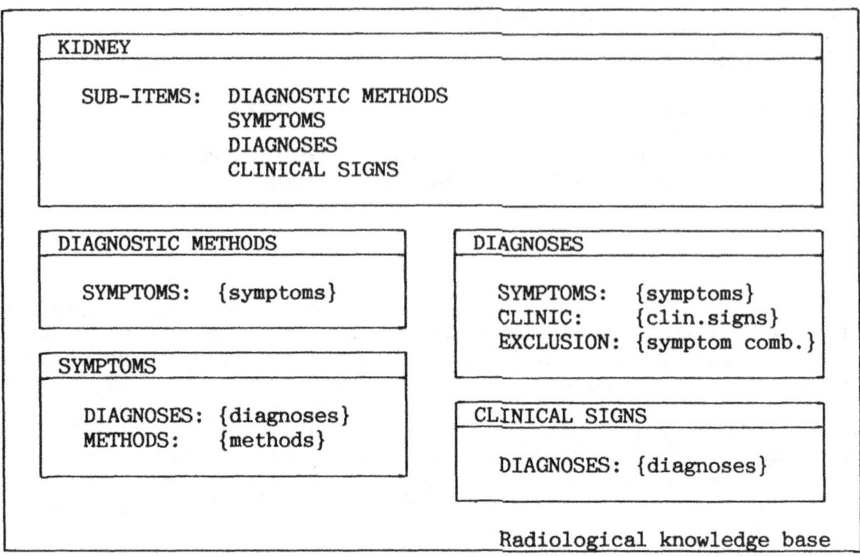

Figure 1: Basic knowledge structure of RADIO.

Knowledge structure of RADIO

The basic structure of the knowledge base is given in Figure 1. Each piece of knowledge is represented in a frame defining one knowledge unit. For instance, the 'DIAGNOSTIC METHODS'-frame has several sub-frames connected to it that represent urography, X-ray, CT, etc. Relations between the knowledge units are defined in terms of slots and slot-values. As an example, the 'METHODS'-slot (defining the relation between a symptom and diagnostic methods) of 'Calcification identical with parenchyma' contains all the methods that symptom may be identified with. Figure 2 gives an example of a that symptom frame.

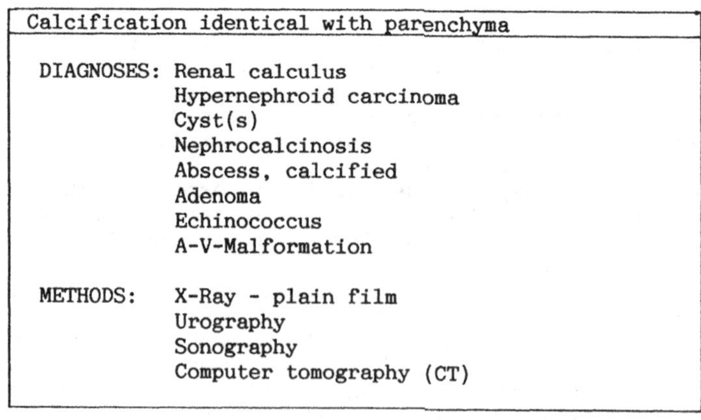

Figure 2: Frame representing the symptom 'Calcification identical with parenchyma'.

The slot-values of the 'DIAGNOSES'-slot of a symptom conform to a an ordering relation in decreasing frequency. The first position ('Renal calculus') gives the most frequent diagnosis. This ordering scheme is used by the consultation module to present an ordered list of possible diagnoses.

Inverse relations are built and maintained automatically. For instance, the 'METHODS'-slot of 'SYMPTOMS' contains an if-added demon, that adds the method in the 'SYMPTOMS'-slot of the diagnostic method, when the method is added to the 'METHODS'-slot of a specific symptom. An equivalent if-removed demon maintains consistency by removing items from the slot-value list. There are demons for all kinds of relations and inverse relations. This guarantees that the same result is achieved regardless if the user edits the 'SYMPTOMS'-slot of a method or the 'METHODS'-slot of a symptom.

Usage of inheritance

Inheritance of slot-values is one of the most essential features of RADIO. Diagnoses and symptoms are hierarchically organized. For symptoms the hierarchy results from the grouping of specialized symptoms to form a class. For diagnoses, the hierarchy is based both, on nosolgical criteria and on considerations about differential diagnosis. Figure 3 shows part of the diagnoses hierarchy.

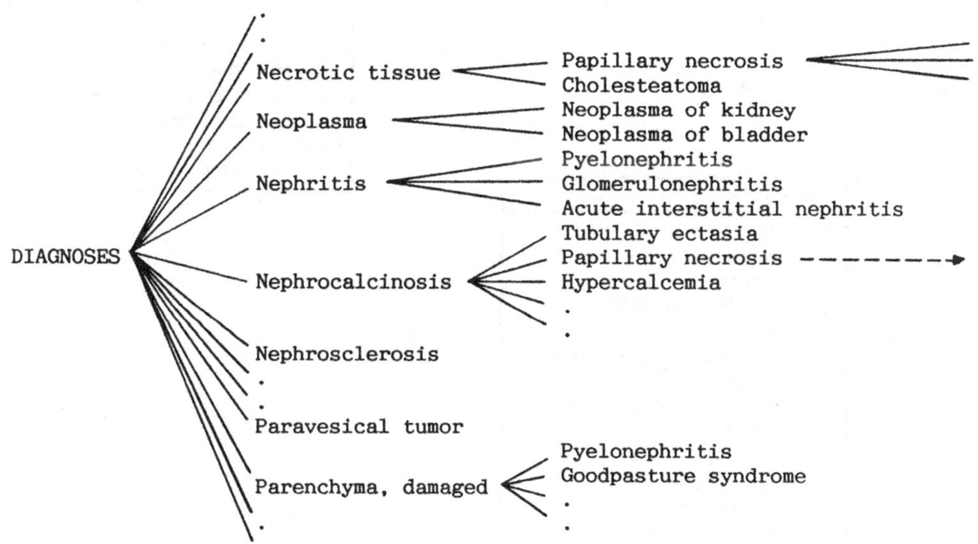

Figure 3: Part of diagnoses hierarchy.

This hierarchy is depicted in form of a tree. But in its true form it defines a net. For instance, 'Pyelonephritis' is a successor of 'Nephritis' and a successor of 'Parenchyma, damaged'. And it is in both cases the same frame. This defines a multiple inheritance scheme. Slot-values will be taken from all the superior frames. Figure 4 gives an example of inheritance of symptoms for the diagnosis 'Pyelonephritis'. The inheritance method used is adding

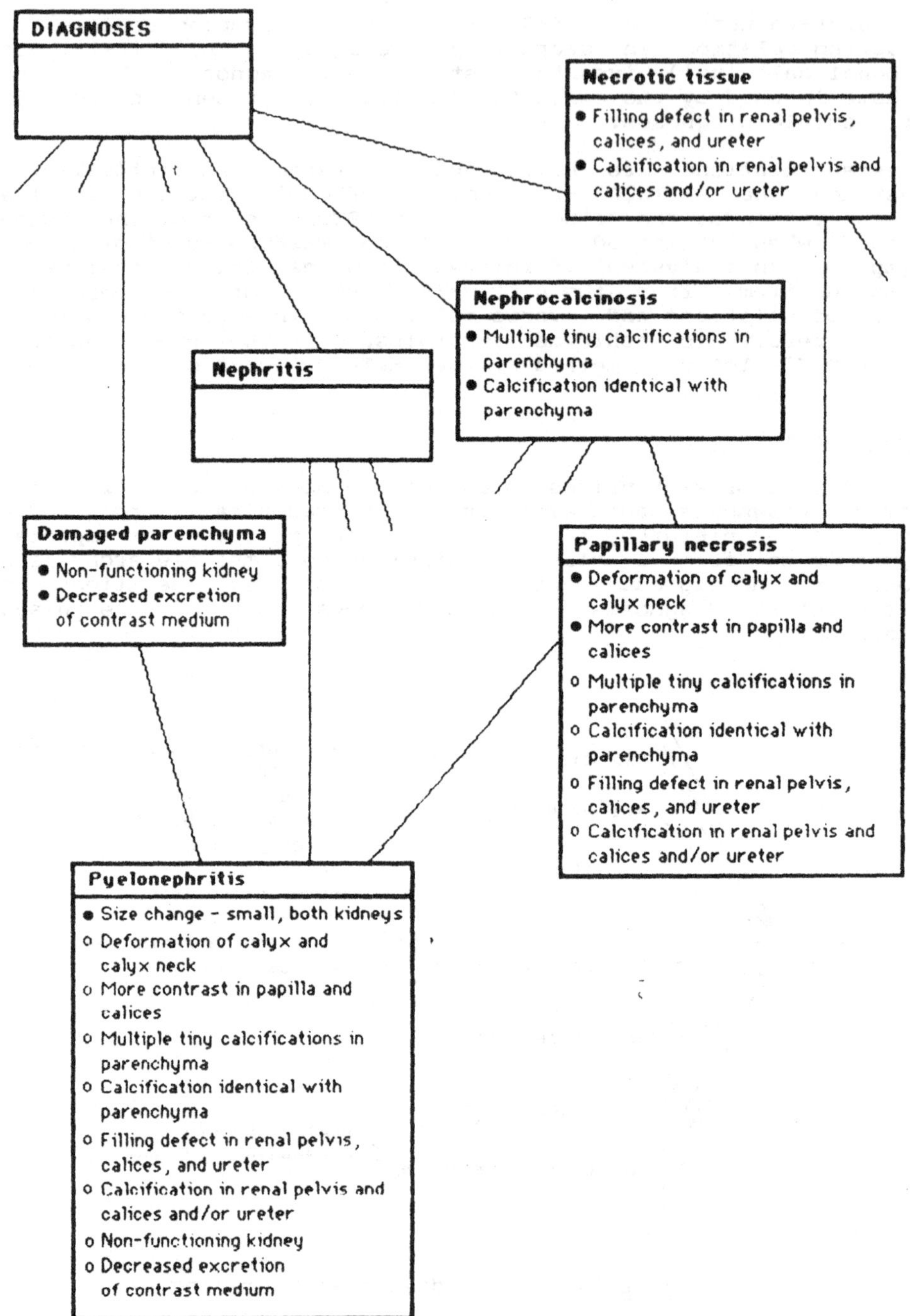

Figure 4: Inheritance of symptoms defined at diagnostic knowledge units (non-inherited symptoms are marked with '●', inherited symptoms are marked with 'o').

the locally defined symptoms and the inherited symptoms. Inherited symptoms are defined by the combination of the symptoms of all superior diagnoses. This cumulating inheritance scheme makes it an easy job to define a more special diagnosis. All that is needed is to add the extra symptoms. On the other hand the knowledge acquisition module has to have a tracing facility that shows where the symptoms are inherited from.

The consultation

A consultation session starts with entering the specific symptoms of the patient. There are two ways to do this:
- select the diagnostic method (e.g.urography) performed. After selection, the system presents a menu containing all the symptoms associated with this method;
- selecting from the full menu of all symptoms.

The selection procedure supports the tree search for symptoms more specific as defined by the hierarchy relation. Symptoms may be selected at any level of detail.

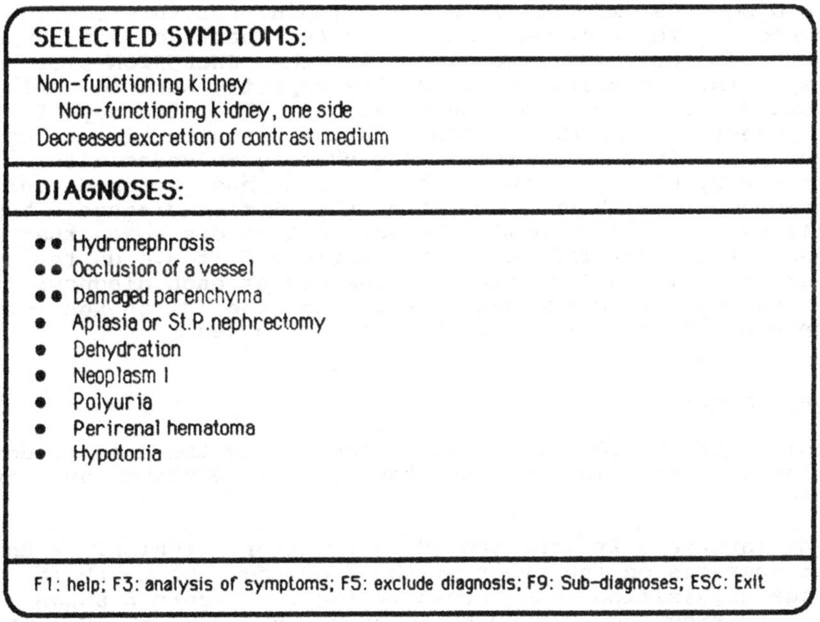

<u>Figure 5</u>: Sample display from a consultation session.

After having entered all known symptoms the possible diagnoses are calculated. Figure 5 gives an example of the display. On top of the display appears a list of the symptoms entered. The hierarchical structure of the symptoms is shown using indentation (for instance, 'Non-functioning kidney, one side' is a sub-item to 'Non-functioning kidney'). All symptoms are used for the generation of diagnosis hypotheses. The generation process is guided by the strategy to find all possible diagnoses. This conforms to the inheritance scheme discussed in the previous chapter. A possible alternative would be to generate from the leaves of the confirmed symptom tree only. This would be a good strategy to find the most plausible diagnosis using the most detailled information.

The list of possible diagnoses (middle of the display) is ordered. First, diagnoses are grouped by the number of supporting symptoms. This gives the number of dots ('●') in front the diagnoses (e.g., 'Hydronephrosis' is supported by two symptoms). Within the groups the ordering is done using the ordering relation of the 'DIAGNOSIS'-slot of the specific symptoms discussed above. The diagnoses list of one symptom is intertwined with the diagnosis list of the second symptom like using a zip-fastener.

To each of the diagnosis presented there are several options:

- analysis of symptoms: this presents the list of all symptoms for the specific diagnosis, with the known symptoms marked. This option can be seen as a first step towards an explanation module, displaying why the diagnosis has been invoked. In addition, future work will concentrate on methods for guiding further symptom collection towards confirmation or rejection of the diagnosis hypothesis;
- excluding the diagnosis: in expert mode the user can exclude the diagnosis if it is not meaningful for this combination of symptoms. The symptom combination (or part of it, upon the user's choice) is recognized in the EXCLUSION-slot of the diagnosis. This option allows the expert user to extend the knowledge base during consultation. The strategy of defining exclude-relations during consultation results in a knowledge base that is able to focus very sharply on cases often seen (as it it very probable that the expert has entered sufficient knowledge). With uncommon cases the list of diagnosis hypothese will be long and will perhaps include diagnoses that are not meaningful. But this seems tolerable and it avoids the undoable task to define all possible exclusions at each diagnosis;
- analysis of sub-diagnoses: following the diagnoses hierarchy the user may look for more specific diseases.

The user interface

The design of RADIO is mainly influenced by the ease of use. The physician will be able to use the system without any training required.

The user interface is implemented in a uniform fashion. A universal browser operates on the items of the knowledge base (both, during knowledge acquisition and consultation). Figure 6 shows a sample display, editing the 'DIAGNOSES'-slot of the symptom 'Calcification identical with parenchyma'). The features of the browser are as follows:
- a border window (status line on top of screen) gives information about the item currently in use;
- the main screen gives a list of sensitive items. The cursor keys are used for item selection (in absence of a mouse). Selected items (i.e., 'Adenoma') are highlighted using the colors of the PC. The browser supports single-selectors (select only one item) and multi-selectors (select a list of items);
- a multiple-page handler operates on lists of items that do not fit on one page;
- a help line (bottom of screen) gives a short information about the possible actions;

```
┌────────────────────────────────────────────────────────────────┐
│ Calcification identical with parenchyma                         │
├────────────────────────────────────────────────────────────────┤
│ Renal calculus                                                  │
│ Hypernephroid carcinoma                                         │
│ Cyst( s )                                                       │
│ Nephrocalcinosis                                                │
│ Abscess, calcified                                              │
│ Adenoma                                                         │
│ Echinococcus                                                    │
│ A-V-Malformation                                                │
│                                                                 │
│                              ┌─────────────────────────────┐    │
│                              │ Menu  (ESC: end)            │    │
│                              ├─────────────────────────────┤    │
│                              │ F2: History of actions      │    │
│                              │ F3: Output element on printer│   │
│                              │ N:  Insert new element      │    │
│                              │ F:  Shift element forward   │    │
│                              │ B:  Shift element backward  │    │
│                              │ V:  View element            │    │
│                              └─────────────────────────────┘    │
├────────────────────────────────────────────────────────────────┤
│ F1: help; N: new diagnosis; F: forward, B: backward shift; ESC: Exit │
└────────────────────────────────────────────────────────────────┘
```

<u>Figure 6</u>: The user interface.

- the browser operates both, menu-driven and command-driven. In
  the menu-driven mode a window (lower right corner) appears
  giving the possible actions to the selected item. An action is
  selected by either selecting the action in the menu window using
  the cursor keys, or by entering the command key;
- the browser may recursively call itself. For instance, the
  'N'-action (adding a new diagnosis) invokes a browser to select
  from the diagnoses hierarchy one element (or define a completely
  new one).

<u>Implementation</u>

RADIO is implemented on an IBM-AT with 640 kB memory and a color
monitor under MS-DOS. It uses VIE-KET as basic knowledge
engineering tool. The current version uses IQLISP as a host
language. Notwithstanding that RADIO runs very fast (besides
garbage collector gaps), we are looking for a compiled AI-language
with run-time support for the PC. The complete knowledge base
(approximately 70 symptoms and 220 diagnoses) is kept in memory.

<u>Summary</u>

A system for radiological diagnostic decision support (RADIO) is
presented, designed and implemented for a personal computer. The
system consists of a consultation module and a knowledge acquistion
module. Both of them are designed for direct use by the
radiological expert on the PC. Design and implementation of RADIO
focus on a stepwise refinement of structure and complexity. After
completion of the first two steps, RADIO utilizes multiple

inheritance of symptoms, diagnoses, diagnostic methods and clinical
signs. It automatically builds inverse relations on knowledge
definition, and it supports the definition of exclusion criteria for
specific diagnoses in an expert user mode during consultation.

The implementation of RADIO concentrates on two goals: on having an
open system that supports future enhancements and structural
changes, and on building a transparent system easy to use even by
the PC non-expert. To cover the first issue, RADIO is implemented
in VIE-KET, our flexible knowledge engineering tool. Ease of use is
provided by a window-oriented management of knowledge during
acquisiton and consultation. To be able to do this, a generic,
extendible browser has been implemented that operates both, command-
and menu-driven.

References

[1] Mars N.J.I., Miller P.L.: Tools for Knowledge Acquisition and
    Verification in Medicine, Proc.10th Annual Symposium on Computer
    Applictions in Medical Care, IEEE, Washington, DC, pp.36-42,
    1986.
[2] Holzbaur C., Pfahringer B.: Synthesis of Hybrid Languages,
    Applied Artificial Intelligence, 1(1)39-52, 1987.
[3] Pfahringer B., Holzbaur C.: VIE-KET User Manual, Austrian
    Research Institute for Artificial Intelligence, Vienna, 1986.
[4] Horn W.: ESDAT - Decision Support for Primary Medical Care, in
    van Bemmel J.H., et al.(eds.), MEDINFO 83, North-Holland,
    Amsterdam, pp.484-487, 1983.
[5] Horn W., Ammer K.: Causal and Anatomical Reasoning in a
    Two-Level Medical Expert System, in O'Shea T.(ed.), Proc.
    ECAI-84, Elsevier, Amsterdam, 1984.
[6] Musen M.A., Fagan L.M., Combs D.M., Shortliffe E.H.:
    Facilitating Knowledge Entry for an Oncology Therapy Advisor
    Using a Model of the Application Area, in Salamon R., et
    al.(eds.), MEDINFO 86, North Holland, Amsterdam, pp.46-50, 1986.
[7] Pokieser H.(ed.): Radiologie - Zielkatalog fuer die Facharzt-
    ausbildung, Oesterreichisches Bundesinstitut fuer Gesundheits-
    wesen, Wien, 1982.

A PC-BASED SHELL FOR CLINICAL INFORMATION SYSTEMS WITH REASONING CAPABILITIES

Fred Wiener and Torgny Groth

Faculty of Medicine, Technion, Israel Institute of Technology,
Haifa, Israel
and
Unit for Biomedical Systems Analysis, Uppsala University,
Uppsala, Sweden

INTRODUCTION

With microcomputers now almost universally available in hospital departments, attention has shifted to the development of the clinical uses of such computers. One area in which much progress has been made is in small to medium scale medical information systems which are designed to log data from a selected group of patients or from a selected portion of the patient record [1]. These data are then used for patient followup, discharge summaries, or clinical research. Another important area which has received much attention is medical expert systems, where expertise in a given specialty is captured by the computer and made available to the medical staff as an electronic consultant [2].

While both types of system have their specific area of application and are indispensible tools for medical researchers, for the bulk of the medical staff these systems seem to be superfluous. Data logging without obtaining feedback and response from the system seems pointless and tedious to the house staff. Expert systems may provide useful advice or diagnoses but the effort required to dialog with the system in order to adequately describe a case will discourage routine use of such systems in clinical situations.

In this paper we describe a shell which combines the advantages of expert systems with the flexibility and ease of access of medical information systems. Data entry can be designed in conformity with clinical procedures and to simulate the forms in routine use. For each data entry session advice and comments can be obtained from the knowledge base. Such reasoning may also be applied to the entire patient record with the appropriate inferences and interpretations presented to the user. This system may provide a vehicle for a more intensive penetration of AI in day to day clinical work.

THE CLINICAL INFORMATION SYSTEM

The clinical information system has a file structure consisting of a main file and any number of subfiles. The main file allows for one record per patient and may consist of patient ID, demographic data and perhaps initial or anamnestic data. The subfiles can contain multiple entries for each patient and are usually limited to a given subject such as laboratory results, followup exams, clinic visits, or progress notes. A patient record in the main file has a pointer to the latest record for that patient in each of the subfiles. The records for a given patient in a subfile are connected by a chain of backward pointers from the latest entry to the first one.

Each record in any file corresponds to a single screen for data input or display. The screen description thus corresponds to the DDL (Data Description Language) in data base systems, except that it also contains information as where each field is to be displayed on the screen and where the cursor is to be positioned for data input. The application designer, who could well be a computer layman, begins by laying out the proposed screen on a 24x80 matrix. The first line contains the screen title. Line 2 is left blank for system messages. The rest of the screen may be used freely and will contain field names and space for data input as well as free text for headings or instructions.

The screen specification is entered into the computer using any available text editor or word processor and stored in a screen file. This file begins with a line giving the screen title and the number of fields present. The field text is a string of up to 30 characters which is sufficient for any field name or heading. Longer texts may be accomodated by placing them in contiguous fields. Next, there is a line for each field giving the field text string, the coordinates (line and column) where the text is to begin, the column (on the same line as the text) for data input, the field length, a single character string ('a', 'i', or 'c') designating the data type, and a number representing the item in the knowledge base with which the data in the field is associated. For fields with no relation to the knowledge base, this number is 0. Fields for headings or instructions will have a line giving the text string and its coordinates, while the data input column, field length, data type, and knowledge base item number all will be 0 or the null string. For example:

```
'FLUID THERAPY - PATIENT ADMISSION' 18
'ID number'            3  1 11 10 'a' 0
'Name'                 3 25 30 20 'a' 0
'Weight, kg'           3 55 66  3 'i' 406
'Arterial BP'          5  4 16  1 'c' 404
'CVP'                  6  4 16  1 'c' 402
'HR'                   7  4 16  3 'i' 403
'Initial Urine, ml/h'  5 37 57  2 'i' 411
'Ongoing WL, dl/d'     6 37 57  2 'i' 407
'Hemoconcentration'    7 37 55  1 'c' 410
'Dehydration signs'            9 25 0 0 '' 0
'Please enter (y/n) next to ' 11 13 0 0 '' 0
'each statement:'             11 40 0 0 '' 0
'dry mucous membranes' 12 23 45 1 'a' 805
'periph Temp normal'   13 23 45 1 'a' 802
'reduced periph circ'  14 23 45 1 'a' 807
'reduced skin turgor'  15 23 45 1 'a' 804
'soft eyeballs'        16 23 45 1 'a' 803
'urine osm, high'      17 23 45 1 'a' 1203
04 arterial BP
severely reduced
reduced
normal
elevated
03 CVP
low
normal
high
03 Hemoconcentration
none
slight
appreciable
```

Three data types are defined: 'a' alphanumeric, 'i' integer (decimal numbers are recorded as alphanumeric data), and 'c' coded. A coded field accepts an integer number which corresponds to a preset text. For each coded field, how many texts there are is stated, followed by the texts themselves. This information is placed in the screen file following the field specifications. A set, respectively, for each field designated as 'c' must appear in the file, as may be seen in the above example. During data input, the user enters the number and the system displays the corresponding text in the data field.

Program DBS is used to display the screen and to check the cursor positions for data entry. This enables the application designer to view the planned screen and to make adjustments in the screen file until the screen is satisfactory. An option in program

DBS prints out the screen together with the coded field definitions. This printout may be used for system documentation and as an aid for data entry. For the screen file shown above, the printout is:

```
                 FLUID THERAPY - PATIENT ADMISSION

ID number              Name                        Weight, kg
   Arterial BP                     Initial Urine, ml/h
   CVP                             Ongoing WL, dl/d
   HR                              Hemoconcentration

                    Dehydration signs

           Please enter (y/n) next to each statement:
                    dry mucous membranes
                    periph Temp normal
                    reduced periph circ
                    reduced skin turgor
                    soft eyeballs
                    urine osm, high


      CODED FIELDS
   arterial BP              CVP                  Hemoconcentration
    1 severely reduced       1 low               1 none
    2 reduced                2 normal            2 slight
    3 normal                 3 high              3 appreciable
    4 elevated
```

After specifying all the required screens, the application designer constructs a menu file consisting of a four-line title for the application, the names of all the screens, and a vector, initially set to zero, giving the number of records recorded in each data file (one file per screen). The system is now ready for data input, which is done using program DB.

Program   DB   begins   by   displaying   the   data   entry   menu:

```
*****************************************************************
*                    UPPSALA UNIVERSITY                        *
*                                                              *
*                    ACADEMIC HOSPITAL                         *
*                                                              *
*                DEPARTMENT of ANESTHESIOLOGY                  *
*                                                              *
*                   FLUID THERAPY ADVISOR                      *
*****************************************************************
```

                         DATA ENTRY

                    1. Patient admission
                    2. Progress notes
                    3. Laboratory results
                         choice

Option 1 must be selected to create a record for the patient in the main file. After the choice is entered, the system displays the screen with the cursor positioned in the first data field which must be the patient ID. The patient ID must be entered as it is the key for data retrieval and for entry of records into the subfiles. If for any other field the data is unknown or irrelevant for the particular case, 'Enter' is pressed and the cursor moved on to the next data field. Exit from the data entry mode is automatic after the last data field or volitional by entering 'x' in the current field. Upon exit, the message

                    'corrections? (y/n)'

is shown on line 2 of the screen. If the reply is 'y', the cursor returns to the first data field. The user paces through the data fields by pressing 'Enter' without entering data. This leaves the current contents of the field intact. When the fields to be corrected are reached, the corrected value is entered. Exit from the corrections mode is also after the last data field or by entering 'x'. Again the 'corrections?' prompt is shown on line 2, and the corrections mode may be entered as often as necessary.

When no more corrections are to be made, 'n' is entered and the prompt 'patient status evaluation? (y/n)' appears on line 2. This is to enable the user to have the newly enterd data evaluated if the knowledge base contains rules for assessing the data's logical consistency or for making recommendations using only a partial description of the case. For the fluid therapy screen shown above, the knowledge base can estimate the patient's level of dehydration as well as his 24 hour fluid needs on the basis of the initial data.

For the main file screen the prompt 'record another patient?' is displayed and as many patients as desired can be entered in one session. When no more patients are to be entered at this time, the response 'n' is entered and the data entery menu reappears.

If data is to be entered in a subfile, the appropriate option is selected and the system asks

'enter patient ID number'

the patient ID is entered and if such a patient is on file, the data entry screen for the selected subfile is displayed. Data entry, corrections, and patient evaluation for the data on the currently entered screen proceeds as described above for the main file, after which the data entry menu again appears. Exit from program DB is by pressing 'Enter' without entering a 'choice'.

Program DBR is used for data retrieval. The data retrieval menu displays the application title and the following options:

DATA RETRIEVAL

1. list of patients
2. display full patient record
3. Patient admission
4. Progress notes
5. Laboratory results
            choice

Since the key to the patient record is the patient ID which may not be known to the user, option 1 displays a list of the names and ID number of all patients in the file. In selecting any one of the other options, the program first asks the user to enter the patient ID. If the patient is on file, the screen or screens corresponding to the selected options are displayed with the data fields filled with the data from the patient's record. For option 2, the full patient record is displayed beginning with the record from the main file and the records from the successive sub-files in the sequence shown in the menu. For each sub-file, all entries for the patient are shown, beginning with the most recent and paging back along the pointer chain through all the patient's records in the sub-file. For all other options only the record or records in the selected file are shown.

When each data filled screen is displayed, the message appears on line 2, 'corrections/update? (y/n)', to allow the user to enter changes into the record. For data such as laboratory findings or intensive care monitoring, the screen may be organized in tabular form, with each succesive column containing data from successive periods of measurement. The tabular form is convenient for following the trend of time dependent data. The fist column of such a screen would be recorded under the data entry program DB. The subsequent columns are recorded using the 'corrections/update' feature of the data retrieval program DBR. Again, exit from the screen during the corrections mode is by entering 'x' in any data field so that the user must advance the cursor to the desired column by successively pressing 'Enter' but can leave the screen immediately after recording the column.

As was the case for data entry, after exit from the corrections mode by replying 'n' to the prompt, the message

'patient status evaluation? (y/n)'

appears and the user may enter 'y' to obtain interpretations and advice from the knowledge base. During data retrieval, however, the data evaluated is cumulative from the first screen shown under the selected option to the screen currently shown. It would seem more relevant to obtain conclusions from the knowledge base only after all the data for the selected option has been accumulated. Therefore, after all the patient's records in the subfile have been retrieved and displayed, the message

'sub-record retrieved patient status evaluation? (y/n)' appears. If the user's selection was one of the subfile options this is the time to answer 'y' to branch to the expert system. If the user had selected the option to 'display full patient record', branching to the expert system should be delayed until the message

'full record retrieved patient status evaluation? (y/n)' is displayed.

A fourth program of the clinical information system, DBQ, provides a facility for formulating queries in each data file and to obtain lists of all patients in the data base who meet the specified conditions. Program DBQ also produces histograms showing all values entered in a given data field and how many cases there are for each value. Another option of the query program builds a sub-data base containig the full records of those cases who satisfy a specific query. This sub-data base can then be querried using program DBQ, thus providing for complex queries which may span several sub-files.

THE EXPERT SHELL

The reasoning capabilities of the system are provided by the expert system shell SMR (Simulating Medical Reasoning). This shell consists of FORTRAN programs which enable an expert to construct his knowledge base and a subroutine for comparing patient data with the rules in the knowledge base to draw the relevant conclusions (i.e, the "inference engine"). The design philosophy, knowledge base structure, and operational procedures of SMR have been described in detail in previous publications [3,4]. Here we limit ourselves to sketching the features of SMR which are needed to understand the interaction with the clinical information system.

An SMR knowledge base consists of a series of rules, each stating a putative inference together with all criteria for confirming or rejecting the inference. Since criteria may include primary observations as well as previously resolved inferences, interconnected rules form a decision pathway. Part of such a pathway in the fluid therapy knowledge base is:

```
rx            Give vasodilator and ionotropic supplement              307
sg  0  33y PCW, high                                                   817
sg  0  17y arterial BP, normal                                        801
sg  0  17y arterial BP, high                                          818
                                                                    -------
rx            Give Diuretic                                            302
sg  0  40y urine output remains low, despite fluid infusion           810
rx  0  10y Give vasodilator and ionotropic supplement                 307
sg  0  10y CVP, normal                                                 813
sg  0  10y arterial BP, normal                                        801
lb 50y  0  Serum potassium < 3.2 mmol/1                               1217
```

These two rules also illustrate the "inference engine" logic. Each criterion is assigned points (those in the third column are for confirmation, those in the second column are for rejection) such that any combination of criteria whose point total is 50 or more is sufficient for resolving the inference. Thus the 'vasodilator and ionotropic supplement ' are given when the PCW is high and the arterial BP is either normal or high. The recommendation to 'Give diuretics' is confirmd if the urine output remains low and any 1 of the other 3 criteria is present. 'Give diuretics' is rejected if the serum potassium is below 3.2 mmol/1, even if the criteria for confirmation are also present.

The rules are entered into the system using a line editor or word processor. The first SMR program compiles the rules, storing the inference and criteria statements from all the rules in a text file, assigning an item number to each statement (the item numbers are shown on the far right in the above illustration). The program constructs a dictionary array for accessing the text file and a decision matrix which is the internal represenatation of the rules.

A second SMR program produces an alphabetized list for each statement type (the statement type must be designated by the expert and placed in the leftmost column of the rules) of all the texts of this type used in the rules. The list gives the item number of the statement followed by its text, referenced to all the rules in which it appears. These lists are then a glossary of the knowledge base's semantics which are defined in SMR as a result of rule formulation. Rules are thus formulated as natural language text organized within a simple, readily understood structure. The item numbers used in the screen file to indicate those data items connected with the kowledge base are obtained from these alphabetized lists.

As may be seen from the rules shown above, there are some criteria which are mutually exclusive and others which the expert may want to associate with a specific numerical measurement. Such criteria groups may be defined using a third program of SMR:

```
      404   arterial BP (1 very low, 2 low, 3 normal, 4 high)
   .0  -   1.0    808     arterial BP, severely reduced
  1.1  -   2.0    806     arterial BP, reduced
  2.1  -   3.0    801     arterial BP, normal
more than 3.0    818     arterial BP, high

      405   increase in hourly urine after infusion, ml
   .0  -  12.0    810     urine output remains low despite fluid infusion
more than 12.0    811     urine output increases after fluid infusion

      411   Initial urine flow, ml/h
   .0  -  15.0   1210     oliguria-anuria <15 ml/hr
 15.1  -  20.0   1207     oligura 15 - 20 ml/hr
 20.1  -  22.0   1208     oligura 20 - 22 ml/hr
 22.1  -  26.0   1205     oligura 22 - 26 ml/hr
 26.1  -  40.0   1202     oligura 26 - 40 ml/hr
more than 40.0    400     Not defined
```

The criteria group name is also assigned an item number which is used in the data description of the information system. Comparing the screen file with the criteria group definitions, we see that 404 in the 'Arterial BP' field and 411 in the 'Initial Urine, ml/h' field relate to the corresponding data groups shown above.

SMR stores patient data in a patient vector, an array containing the item numbers and data group values relevant to the patient's condition. In a stand alone application of SMR, data is inserted into the patient vector by replying to data gathering protocols and questionnaires designed by the expert. In the present system these data are automatically inserted into the patient vector by the information system programs. Both DB and DBR, when accessing the screen file in receiving or displaying data check if the last number in the field description line is > 0, as for example:

```
'Arterial BP'          5 4 16  1 'c'  404
'Initial Urine, ml/h'  5 37 57 2 'i'  411
'urine osm, high'      17 23 45 1 'a' 1203
```

For the field type 'a', if the data is 'y' the item number is recorded in the patient vector, if it is 'n', the item number prefixed by '-' is recorded. For a numeric field, 'c' multiple choice or 'i' integer, a composite of the criteria group number and the data value is constructed in comformity with SMR usage (e.g., low BP is 4001, 25 ml/h Initial Urine is 11025) and inserted in the patient vector.

As stated above program DB constructs a patient vector from one screen only, while program DBR accumulates data from all screens displayed for the selected option. If the user reply to the 'patient status evaluation?' prompt is 'y', the patient vector is passed as a parameter to the SMR inference engine subroutine. The subroutine examines the decision matrix line by line. For each rule, if the criteria are present in the patient vector, points are accumulated for confirmation or rejection. The item number of a confirmed inference is appended to the patient vector. Previous conclusions are thus available as criteria in examining subsequent rules. At the end of this process the patient status report is displayed:

Patient #27750

         clinical findings
  56 body mass, kg
   3 arterial BP (1 very low, 2 low, 3 normal, 4 high)
   3 CVP (1 low, 2 normal, 3 high)
  23 urine flow, ml/h
  15 ongoing water loss,dl/d
   2 hemoconcentration (1 none, 2 slight, 3 appreciable)
y dry mucous membranes
n peripheral temperature, normal
y skin turgor, reduced
y urine osmolarity, high

         subsequent findings
 132 serum sodium, mmol/l
  33 serum potassium, mmol/l*10

         subsequent findings
   2 PCW pressure (1-low 2-high)
y Fluids for hour 1 administered
  30 subsequent urine output, ml/h

         subsequent findings
y Fluids for hour 2-3 administered
  34 subsequent urine output, ml/h
  11 increase in hourly urine after infusion

based on your findings, the inferences are
      HOUR 4-7, Ringer-acetat, 20% of 24-hr water need
      add (135-Na)*.0095*body weight mmol/hr Sodium to
          to ongoing infusion
      add 5 mmol/hr K *(wt, kg/75) to ongoing infusion

FLUID PRESCRIPTION
  5980 ml, 24 hour water requirement
     Hour 4-7:   299 ml/h, Ringer-acetat infusion (1196 ml for period)
                 1.6 mmol/h Na supplm   5.3 mmol Na per liter bag
                 3.7 mmol/h K supplm   12.5 mmol K per liter bag

measures for patient management:
      Give vasodilator and ionotropic supplement
      Give Diuretic
      test electrolytes

for followup, consider items listed under each statement
  20 HOUR 8-15, Glucose 100, 20% of 24-hr water need
        50 Fluids for hour 4-7 administered

The report is divided into 3 sections, the patient data as recorded in the patient vector, the confirmed conclusions from the knowledge base, and the data to be obtained on follow-up. The first part of the patient data section, labeled 'clinical findings', is the data from the patient's record in the main file of the information system. The parts labelled 'subsequent findings' correspond to each record for the patient in each of the subfiles. The records from the subfiles are displayed by SMR from the earliest to the most recent, which is the reverse of the sequence of screens in the information system. The value for 'increase in hourly urine after infusion' is calculated by SMR from the data in the files.

The recommendations from the knowledge base starts with the fluid therapy inferences. This is followed by the instructions to the nurse labelled 'FLUID PRESCRIPTION' which is calculated by a 'special applications' subroutine [4] of SMR. The other rules confirmed from the knowledge base are labelled 'measures for patient management'. Note that the vasodilator and diuretics rules illustrated above were confirmed for this case.

The report ends with a listing of the activated rules (which have accumulated 20 or more points) and the data required to confirm them. In the case shown, the fluid therapy step following the one currently recommended has been activated and will be confirmed when the completion of the current step is recorded in the information system.

SUMMARY

We have shown in this paper the structure and operation of a simple but effective clinical information system; how it is interfaced with a rule based inference system to the enhancement of both systems. The combined package provides a practical method for introducing AI into routine clinical practice. The application we have chosen to illustrate this package is one of several projects now underway at Uppsala University Hospital for developing user-friendly knowledge based systems for patient management and follow-up.

REFERENCES

1. Stead WW, Garret LE Jr, and Hammon WE: Practicing nephrology
   with a computerized medical record.
   Kidney Int 124:446-454 (1983).

2. Wiener F, Fayman M, Teitelman U, and Bursztein S: Computerized
   medical reasoning in diagnosis and treatment of acid-base
   disorders. Crit. Care Med. 11, 470-475 (1983)

3. Wiener F: SMR (Simulating Medical Reasoning): An expert shell
   for non-AI experts. Comput. Programs in Biomed. (In press).

4. Wiener F and Groth F: A system for simulating medical reasoning
   (SMR) providing expertise in clical computer applications.
   Automedica (In press).

The kernel mechanism
for handling assumptions and justifications
and its application to the biotechnologies

Cherubini, M.A. [*] ; Cerri,S.A. [* °] ; Sbarbati,R. [§]

[*] Knowledge Engineering Research Unit, Mario Negri Institute for Pharmacological Research, Via Eritrea, 62, 20157 MILANO, Italy

[°] Dipartimento di Scienze dell'Informazione, Università di Milano, Via Moretto da Brescia, 9; 20133 MILANO, Italy

[§] C.N.R. Institute of Clinical Physiology, Via Savi, 8; 56100 PISA, Italy

0. ABSTRACT

In fields such as medicine and biology knowledge is still empirical and in rapid development. It is therefore needed to develop theories and systems for handling incomplete -thus possibly contradictory- knowledge, in order to be able to incorporate new knowledge that arises from the interpretation of the experimental outcomes and to acquire knowledge by communicating with other (human or artificial) experts.

The paper presents the kernel of a set of possible knowledge-based systems that integrate assumptions and justifications for the interpretation of new knowledge in order to control the possible non-monotonic reasoning processes.

The behaviour of the system is described by means of examples taken by an advisor for decision making in performing laboratory experiments for the production of monoclonal antibodies which is currently under study.

1. INTRODUCTION

The preliminary architecture of a skilled, artificial assistant to the massive production of monoclonal antibodies [Milstein '80] has been presented in [Cerri et al. '85]. The applicative domain has required not only to make an adequate use of traditional A.I. techniques, but also to develop new techniques beyond the current state of the art.

Among other not yet fully solved, fundamental problems, we will discuss here the particular one due to the need of updating assumptions in an experimental and cooperative problem solving activity done by agents that do not have a complete access to each other's available knowledge.

This process of knowledge revision is common in medical and biological decision making where most knowledge is still empirical and in rapid development : the domain expert must confront him/herself with new findings and observations obtained both from his/her own experiments and from those carried out in other laboratories.

Human decision making is conditionated by the limits intrinsic to   human reasoning: often one is forced by many factors to take decisions for which there is not any proof or certain evidence, but which at the moment seem the most plausible. The conclusions that are derived from these decisions could eventually prove to be contradictory with respect to new information or with respect to facts obtained from subsequent deductions. This implies the need of retracting the decisions taken in order to eliminate the contradictions. This kind of non-monotonic reasoning is based not only on facts, but also on assumptions that could be abandoned in the light of new findings.

Reasoning programs usually construct (computational) models of situations. In order to make sure that these models are consistent with the introduction of new information as well as with possible unforeseen changes in the situations that are being modelled, the programs must often eliminate or change parts of their models. These changes may themselves produce further changes because certain parts of the model are constructed starting from other parts of the model and therefore the former parts assume the validity of the latter.

We are thus confronted with the problem of making changes in the computational models, and most importantly with the problem of constructing models flexible enough to be modified continuously with ease.

RETRACTION OF ASSUMPTIONS DUE TO INCONSISTENCIES WITH EXPERIMENTAL OUTCOMES

Let us consider, for example, during the cloning and screening phases [Milstein '80], the real possibility for the hybridomas to loose the capacity of synthetizing specific antibodies [see also: Cerri et al. '85]. In fact one may often observe in the first screening phases a reaction of the antigen to certain antibodies, while this is absent in successive screening phases .

The correct retraction of assumptions implies an explicit representation of knowledge that clarifies the dependencies among different data. Each fact, therefore, must be associated with a justification which describes how that fact was deduced from antecedent facts. In the presence of a contradiction it is possible to go back, through justifications, to the facts that caused it.

RETRACTION OF ASSUMPTIONS DUE TO INCONSISTENCIES WITH INFORMATION FROM EXTERNAL SOURCES

The possibility of acquiring, from an external source, new knowledge that is not necessarily reliable, (in our example, for instance : knowledge regarding new cell banks, new methods or new theories) fosters the need for a "sensitive" system able to *critically* incorporate knowledge. This requires to evaluate the consistency of the knowledge in the system with regard to that acquired from outside.

Each fact must therefore be associated not only to a justification but also to a set of assumptions which constitute the hypoteses under which it holds. By assumptions we mean the hypotheses formed by the expert and not the data assumed to be true for the experiment continuation or for the problem's solution.

Therefore, **while a justification represents the set of facts and rules used for the deduction of a datum, the assumptions constitute the** context (in our case: the laboratory context) under which a datum is believed to be valid.

2. PREVIOUS RESEARCH RELEVANT TO OURS

Studies dealing with these two concepts, i.e. assumptions and justifications, have been undertaken by several groups. Some of these studies [Doyle '79, McAllester '80] tend to solve the problem of the consistency of a single knowledge base using the mechanism of the justifications associated to each datum. Other studies [de Kleer '86, Martins and Shapiro '83] extend the solution by dealing with the coexistence of more than one, possibly contradictory knowledge bases and allowing an interaction among these knowledge-bases through the mechanism of assumptions associated to each datum.

The fundamental difference between justification-based and assumption-based non monotonic reasoning systems is the following. The <u>manipulation of justifications</u> -as in Doyle's and McAllester's Truth Maintenance Systems- enables the identification of a single solution maintaining the consistency of a single database (there is a single context), while the <u>manipulation of assumptions</u> -as in deKleer and Martin's systems- enables to control the entire research space without the restriction that the knowledge base must be consistent, allows the possibility for multiple context to coexist, and facilitates the transfer of information from one point in the space to another.

Our kernel system integrates the justification and the assumption-based techniques into a single control mechanism called the context manager.

A mechanism for controlling ambiguity by allowing multiple contexts to be simultaneously active within a single description has also been reported in [Cerri, 1986]. The conclusions of that paper suggest the integration of context-dependency within the control primitives of the knowledge representation language KRS.

3. THE MANAGEMENT OF CONTEXTS

The kernel mechanism that we have developed allows to maintain the consistency of a knowledge base in both situations when the user (i.e. the expert) does not posses sufficient information while solving his/her own problem and when information from other contexts has to be introduced into the system. Both mechanism of manipulating justifications and assumptions have been used.

The system is implemented in FranzLisp / MRS [Genesereth et al. '84] and extensively documented in [Cherubini '87].

It consists of three components: a Problem Solver, a Truth Maintenance System and a Context Manager (Fig. 0).

The Problem Solver (of the first order) includes the domain knowledge (facts and rules), a control strategy, i. e. the criterion about which rule or knowledge has to be used, and a specific inference mechanism : forward chaining. The Problem Solver makes inferences from known and deduced facts and each deduced fact is comunicated to the TMS.

The TMS mantains the consistency of the current knowledge base (associated to the laboratory) as well as after the acquisition of knowledge from other knowledge bases (associated to other laboratories). The mechanism of dependency-directed backtracking [Stallman and Sussman '77] is used when a contradiction occurs. This mechanism (possibly) eliminates knowledge from the database in order to restore the consistency. The assertions responsible for the inconsistency are retracted independently from the order of their generation.The TMS is also able to give the user the reason of its own conclusions.

The Context Manager (CM) stores the laboratory hypotheses and believes. By interacting with the TMS and the Problem Solver, the CM is able to evaluate and manage the information obtained from other laboratories.

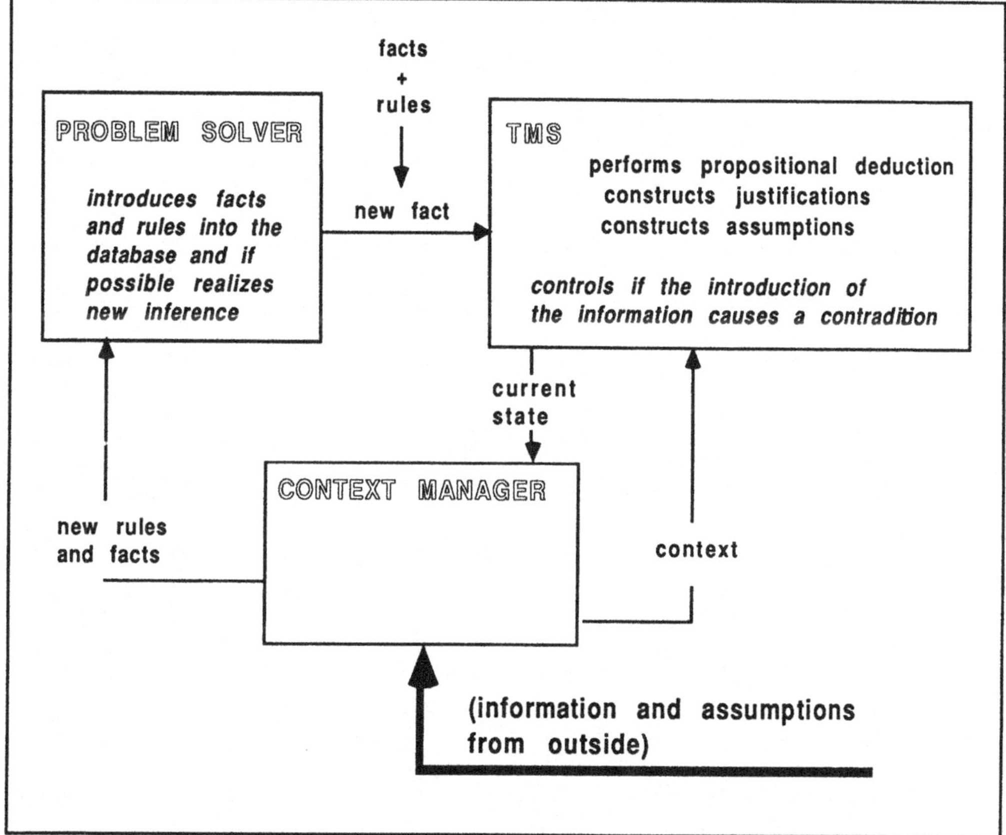

Fig. 0 : The architecture of the kernel

The rationale is that to each Problem Solver datum is associated both a justification which specify how the datum depends from other data and a set of assumptions which are the hypoteses under which the datum holds. The set of assumptions describe **the laboratory context**, on which the CM operates to make possible the information exchange with other Context Managers. An information is accepted only if the associated assumptions are a subset of the assumptions of the evaluation context, or if their introduction in the knowledge base does not cause contradiction in the same context.

4. THE INTERACTION OF CONTEXT MANAGERS

In the following examples different laboratories need an exchange of information. The interaction highlights the functionalities supported by our system and should clarify both its specific application and the possible generalizations.

Example 1

Laboratories A and B cooperate for the production of a monoclonal antibody

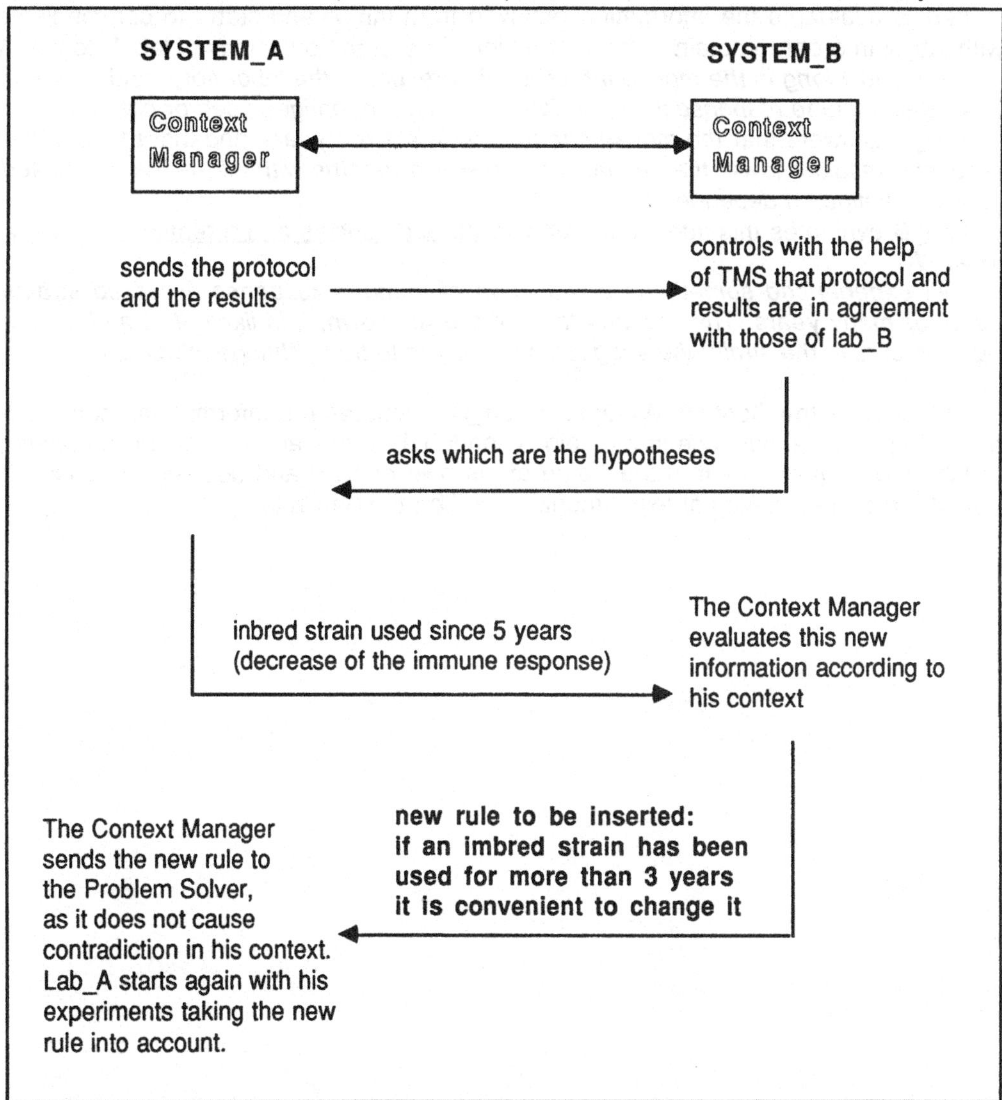

Fig. 1 : The acquisition of a new rule from outside (example 1).

against endothelial cells. Both follow similar protocols, but lab_B has more experience than lab_A in the field.

Researchers in lab_A start by immunizing mice with endothelial cells (the antigen). At time t the animal serum shows - surprisingly - an antibody concentration lower than usal. In spite of that, the researchers in lab_A decide to proceed to the fusion phase, i. e., they fuse B-lymfocytes with mieloma cells to produce hybridomas. Hybridomas are then screened to test their positivity for the antigen. Only one hybridoma results to be positive and this positivity is lost in the successive screenings.

Lab_A then calls lab_B to look for a solution to his problems and sends a protocol of the experiment together with a file of the results (Fig. 1).
Lab_B evaluates the information received from lab_A and starts to communicate with lab_A in order to obtain more information. One of the questions of lab_B to lab_A is : *since how long is the mouse inbred strain present in the laboratory* and *whether this weak immune response is dependent of the type of antigen used or not..*
Lab_A answers that *the mouse strain is there since 5 years* and that *the immune response resulted lower than usual in the last two months with all the antigens used (i. e. it is antigen-independent).*
Lab_B evaluates this information and sends <u>a diagnosis and a remedial</u> (therapy) to lab_A:
"we experienced sometimes a reduction of immune response in inbred strains used for many years. The reasons for that are unknown; it is likely that an immune aging occurs in the strain: the suggested therapy is to try to change inbred strain".

In this case the Context Manager of lab_B evaluates the informations sent by A according to his experience in the field. Then lab_B sends an information (a therapy) to lab_A who evaluates it in the frame of his own context and decides to acquire it. Lab_A start a new series of experiments based on the new rule.

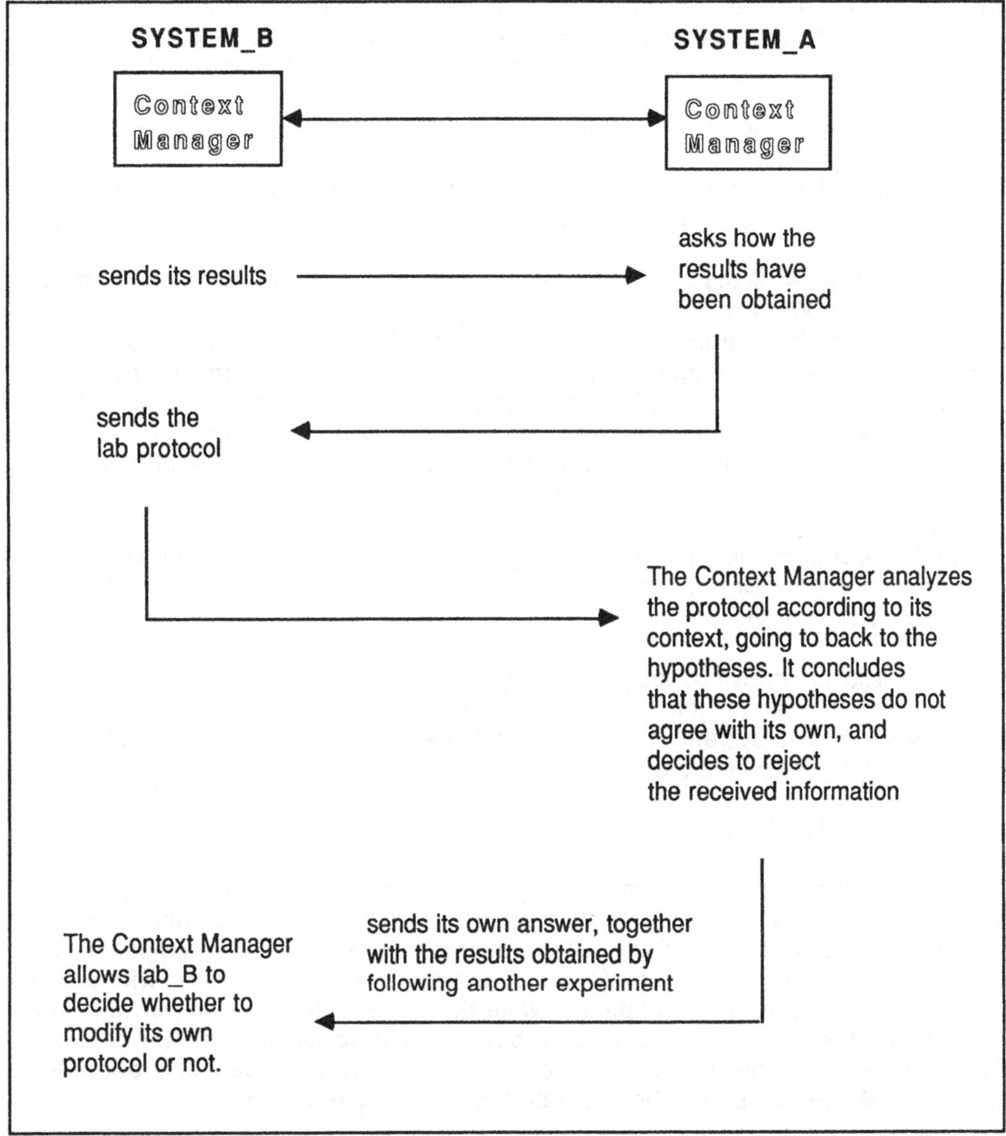

Fig. 2 : Two laboratories operate under different assumptions (example 2).

Example 2

Two laboratories A and B decide to work to the production of MoAb's to a cellular viral component (antigen) particulary difficult to be purified, but extremely important for the diagnosis of a new desease.

The immunogen is a crude substance which evoques response also against unwanted antigen. One consequence is that the screening procedure to isolate the MoAb's will be long and laborious, unless a suited fusion procedure is chosen.

Lab_B follows the protocol of fusion in PEG [Milstein '80] traditionally used with good results.

Lab_A decides instead to follow the new, more selective technique [Lo et al. '84] of

avidin-biotin fusion in electric field, experimented a few times with excellent results. The cloning and screening phases are performed according to a traditional protocol.

Lab_B obtains 40% of positive clones and considers it as a good result. At this time lab_B comes to know about the effort of lab_A to produce MoAb's to same antigen and want to start a confrontation with lab_A, under the agreement that lab_A will not utilize its results without permission. Lab_B then sends all the informations to lab_A.

Lab_A, which had obtained 80% of positive clones for the same antigen, receives the information from B, compares the two results and draws the conclusion that the start hypotheses of the two laboratories are different. The experts are charged of the further decision which of the two protocol is more adequate (Fig. 3).

In this second example there is diversity between two theories (the hypotheses are different). In particular system_B refers to a fusion procedure different from that preferred by system_A.

The conclusion drawn by lab_A is that the two theories are different and it decides to send this conclusion together with its own results and protocol to system_B. In spite of the better results system_A does not decide which is the most adequate protocol and leaves this decision to the experts.

5. CONCLUSIONS

From the challenging problem of designing an advisor assisting in taking decisions about how to operate in the laboratory for the production of monoclonal antibodies, we have considered the specific issues related to the retraction of assumptions due to new experimental findings or to information communicated from outside where similar -but not equal- decisions are taken. While the current literature on non monotonic reasoning either considers justification-based backtracking in a single context, or assumption-based multiple context reasoning, we have noticed that our advisor requires both mechanisms to be integrated into an unique system.

This new kernel consistency checker has been designed and developed so that it may now be embedded into any knowledge-based system that has similar characteristics. It mantains the consistency within each knowledge base by handling justifications : when a contradiction occurs, it activates a traditional dependency directed backtracking module. On the contrary, when new information coming from "outside" (experimental outcomes or messages from other experts) is in contradiction with the system's knowledge, a module is activated that manipulates the assumptions associated to facts and rules in the context.

The detailed analysis of two stereotypical problems in the application domain of production of antibodies has confirmed our intuition about the adequacy of the system we have designed to a large set of situations occurring in the domain of biomedical decision making.

6. REFERENCES

[Cerri et al. '85]
Cerri, S. A., Cherubini, M. A., Minciotti, C., Saffiotti, A. and Sbarbati, R.
Towards a skilled assistant in the industrial production of monoclonal antibodies, In: Artificial Intelligence in Medicine, De Lotto, I. and Stefanelli, M. (eds.). North Holland : 95-105 (1985) .

[Cerri '86]
Cerri,S.A.
Ambiguity in knowledge representation.
In :   Proc. 7th Eur.Conf. on Artificial Intelligence ECAI-86, Brighton; Vol. II: 182-187; North Holland, 1986

[Cherubini '87]
Cherubini, M. A.
Ragionamento con informazione incompleta (Reasoning with Incomplete Knowledge) Thesis, Department of Informatics, Univ. of Pisa, Feb. 1987 (in Italian)

[de Kleer '86]
de Kleer, J.
An Assumption-based TMS . Artificial Intelligence , 28 : 127-162 (1986)

[Doyle '79]
Doyle, J.
A Truth Maintenance System . Artificial Intelligence 12 : 231-272 (1979).

[Genesereth et al. '84]
Genesereth, M. R., Greiner, R., Grinberg, M. R. and Smith, D. E.
The MRS dictionary . Stanford Heuristic Programming Project, Report HPP-80-24 (1984).

[Lo et al. '84]
Lo, M. M. S., Tsong, T. Y., Conrad, M. K., Strittmatter, S. M., Hester, L. D. and Snyder, S. H.
Monoclonal antibody production by receptor-mediated electrically induced cell fusion .
Nature 310 : 792-794 (1984) .

[Martins and Shapiro '83]
Martins,J.P., Shapiro,S.C.
Reasoning in Multiple Belief Space.Proc. 8th IJCAI: 370-373 (1983)

[McAllester '80]
McAllester, D. A.
An Outlook on Truth Maintenance. MIT AI-Memo 551 (1980) .

[Milstein '80]
Milstein, C.
Monoclonal antibodies. Scientific American 243 : 4-56 (1980).

[Stallman and Sussman '77]
Stallman,R.M. and Sussman, G. J.
Forward reasoning and dependency-directed-backtracking in a system for computer-aided circuit analysis. Artificial Intelligence 9 : 135-196 (1977).

General Session

MAN-MACHINE INTERACTION IN CHECK

Luca CONSOLE, Mauro FOSSA, Pietro TORASSO
Dipartimento di Informatica - Universita' di Torino, TORINO (Italy)

Gianpaolo MOLINO
Dipartimento di Biomedicina - Universita' di Torino, TORINO (Italy)

Cesare CRAVETTO
Ospedale S.Giovanni Battista e Citta' di Torino, TORINO (Italy)

ABSTRACT. The paper presents an "ideal scenario" for man-machine interaction in diagnostic expert systems in medicine and discusses how and at what extent the interface of CHECK obeys the requirements. The architecture of CHECK (a diagnostic system in the field of hepatology, able to reason both at heuristic level and at causal one) is sketched. The acquisition of data has been subdivided into "receptive", "evocative" and "exploratory" phases, whereas the explanation mechanism exploits the causal knowledge. The advantages of a graphical interface are also discussed.

1.INTRODUCTION

The design of the man-machine interface is one of the critical points in the development of a medical diagnostic expert system. In fact the acceptability of the system in the medical environment strongly depends on the way it interacts with users (physicians)[1]. This is certainly true independently on the actual application of the system, e.g. for clinical or tutoring purposes [2][3].
With the term "man-machine interface" one generally intends to enumerate several different forms of interaction, e.g. data acquisition or explanation (some authors use the term 'knowledge acquisition' for the phase of acquisition of data from the user in an expert system consultation; We prefer to reserve this term to the design of a knowledge base by a knowledge engineer). The need for flexible and friendly interfaces has been felt as a problem immediately after the development of the first prototype systems (e.g one of the prerequisites settled at the beginning of the MYCIN project was the ability of the system to "understand questions, acquire new knowledge, explain its advice" [4] [5]). In all the successive diagnostic expert systems great attention has been posed to the design of interfaces and in particular to the construction of powerful explanation modules.

Researchers have studied the possibility of building natural language interfaces for expert systems with the aim of allowing the user to introduce data and receive explanations in an easy to understand way (at present natural language interfaces are being developed as research prototypes, for example see [6]). An alternative that has been widely adopted in many recent systems is the construction of graphical interfaces, which are argued as the most natural and easy to use for people without informatic background.
In particular different types of graphical interfaces have been proposed, e.g. OPAL [7] provides the user of ONCOCIN [8], an expert system for assistance in cancer treatment, with forms to fill with the chosen therapy; GUIDON-WATCH [9] provides a multi-windows interface for both a consultation system (HERACLES [10]) and a tutoring system (GUIDON2); in ESDAT [11] an interaction system based on a touch-sensitive screen has been developed; Xi [12] provides the user with a window and menu oriented consultation interface.

---

* The research described in this paper was partially supported by grants from CNR (Strategic Project on Expert Systems in Medicine), MPI and Regione Piemonte.

It is worth noting that in these systems both the problems of which pieces of information should be asked or presented to the user and how these tasks should be performed have been considered [9]. In particular the most critical problem is the first one, i.e. deciding which pieces of information should be asked (explained) and when; the choice and design of a particular form of interaction is far less critical.
A unifying aspect of such researches is the great effort to provide physicians with expert systems which are powerful and accurate on the diagnostic point of view, easy to use, possibly in a very natural way and able to explain/justify precisely their advices.

In the following we will present an ideal scenario for the interactions between physicians and diagnostic expert systems, the influences of this scenario on the architecture of the system itself an we will finally discuss how we have designed the interface module of CHECK [13], a two levels diagnostic expert systems, basing on such scenario.

2. MAN-MACHINE INTERFACE: AN IDEAL SCENARIO

In this section we briefly describe which are the points to be considered when designing a man-machine interface for a medical diagnostic expert system and we present a general scheme of architecture that addresses such problems. The architecture is not bound to a particular implementation or to particular facilities, i.e. it can be practically built in different forms. The ideas we discuss derive from a precise analysis of the behaviour of a physician when he/she visits a patient and from considering how a system should look like and behave to be accepted in a medical environment.

Classically a distinction is made between interfaces for data acquisition and for explanation; we will partially follow this distinction even if, actually, there is a strict correlation between the two phases, as it will be clearer later.

When considering the phase of data acquisition the typical problems to be solved are:
- Should the system ask data or should the user introduce them voluntarily?
- When and how should data be introduced by the user?
- When and how should data be asked by the system?
- How the request of data is connected to the "reasoning phase" of the system?
A possible way to answer these questions is to analyse the behaviour of a physician when he/she visits a patient and try to simulate it. In a first phase the attitude of the physician is mainly *receptive*, that is initial data are provided by the patient (anamnestic phase guided by the patient). The acquisition of initial data may suggest some "*ideas*" about the status of the patient, which in turn may lead the physician to ask some more specific questions to the patient to confirm or reject such ideas (anamnestic phase guided by the physician or *evocative phase*). Initial (general) diagnostic hypotheses are then built through processes of unification and/or differentiation between "ideas" and new (more specific) data (physical examination, laboratory data) are requested to confirm and, possibly, specialise such hypotheses (*exploratory phase*).

This brief analysis suggest the main lines of design of a data acquisition module. The module should be characterised by two different phases:
- a passive phase;
- an active phase.
The first one corresponds to the initial data acquisition from the patient, the second to the phases in which the physician asks data and it is therefore strictly connected and interrelated with the "reasoning process" of the system, in the sense that data acquisition both guides and is guided by the reasoning process (evocative and exploratory phases discussed above). It should be clear that such a general scheme can be put in practice in different ways: for example in the passive phase data could be introduced by the user in natural language or through a graphical interface. What it is worth noting is that the design of the data acquisition module cannot be made independently on the design of the problem solving architecture of the system,

since they are strictly correlated.

A second problem to be considered concerns the explanation components of the system. The ability to produce precise explanations, in terms that are familiar to the physician, is mandatory for a system could be accepted in a clinical (tutoring) context. Such explanations have to be tuned, considering the specific use of the system (e.g. the kind of explanations to be given to a student using the system for tutorial purposes must be different from that given to a physician, using the system as a clinical consultant).
If one considers first generation (heuristic) diagnostic expert systems, i.e. most of the systems developed so far, one will discover that their explanation component is too weak for an actual and successful use of the system in a medical context. Such explanations, whose specific form depends on the knowledge representation formalism adopted (e.g. production rules or frames) are very poor (e.g. the set of applied rules or the instantiated copy of a frame) and not natural, or even easy to understand for a physician. They can hardly convince a physician of the correctness of a diagnosis, they certainly cannot provide the deep motivations of an advice or be very useful in a tutoring context.

It has been recently suggested [14][15][16] that a correct way to build powerful and easy to understand and accept explanations is to rely on different forms of knowledge than heuristic one. In particular the adoption of deep causal/temporal models can be useful to provide the user with proper pathophysiological explanations. The combination of heuristic and deep knowledge corresponds to the situation of a physician which uses both relationships deriving from experience (heuristic knowledge) and pathophysiological knowledge: the former can produce diagnostic hypotheses in many cases, the latter can be used to solve hard or unusual cases or to deeply confirm and explain hypotheses suggested by the experience.
Moreover the adoption of deep models can be useful to produce general explanations about a particular domain [13]. In this way the user can inquire the system independently from the data of a particular case, and get information about the possible causes (consequences) of a disease (pathophysiological state or manifestation). Such forms of explanation can be useful for tutoring purposes.

In the following, after a brief outline of CHECK (for more details see [13][17]), we shall discuss how in this system the ideal scenario for man-machine interface described above has been put in practice.

3. THE CHECK SYSTEM

CHECK (Combining HEuristic and Causal Knowledge) is a two levels diagnostic expert system based on the combination of heuristic and deep models of a specific domain.
In the heuristic level knowledge is represented by means of a hybrid formalism, combining frames and production rules, derived from LITO2, a heuristic expert system we designed in the last years (for a description of LITO2 see [18][19]). In this scheme heuristic relationships between data (findings) and hypotheses are represented, with the aim of achieving diagnostic hypotheses easily. The frames are hierarchically connected, so that coarse diagnostic hypotheses and specific diseases are represented; moreover other kinds of links are used to define associational/alternative relationships between the hypotheses themselves.
In the deep level of CHECK pathophysiological knowledge is represented through the use of causal networks. In such networks causal (temporal) relationships between pathophysiological states, their manifestations and diagnostic hypotheses are represented.

Before describing in more details each one of the two levels, let us briefly discuss how they cooperate during the diagnostic process.
A first point to notice is that the database, which is shared by the two levels, is structured and hierarchically organised, that is, each datum is represented by a frame-like structure and classes of data and hierarchies of classes are defined. Moreover multiple views of the data hierarchy are provided, to be used at different steps during the reasoning process (see [13] for more details). The importance of this organisation will be clear later, when the data acquisition interface will be described.

To understand how control is passed between the two levels it is worth noting that there is a strict correspondence between the frames of the surface level and a particular kind of nodes of the deep one, the HYPOTHESIS nodes: to each frame correspond one hypothesis node in the causal network. During a consultation surface knowledge is invoked first, that is reasoning in the frame system is performed and a hypothetical solution generated. At this time the deep causal level is invoked to confirm and explain the diagnosis, that is a portion of the global network which can confirm and explain the diagnosis from its ultimate causes to its final manifestation is searched and instantiated, starting from the corresponding HYPOTHESIS node. If the hypothesis can be confirmed, at first a complete and precise explanation is generated, then unaccounted and/or unexpected data are taken into account and correlated hypotheses suggested (to the surface level or to the user, depending on their nature). In this way multiple disease solutions can be generated. If the hypothesis is rejected, alternative hypotheses to be considered are suggested to the surface level and control is passed to it.

4. MAN-MACHINE INTERFACE IN CHECK

In this section we discuss how the man-machine interface of CHECK has been designed, having as a model the ideal scenario described in section 2. In particular we have chosen to develop a graphical interface, since we believe that, at present, this is the best choice to build systems that are easy and natural to use for a novice. In the following we will distinguish between the phases of data acquisition and explanation, just to make the description clearer.

4.1 Data Acquisition

The major part of the effort for the acquisition of data from the user is charged to the heuristic module of CHECK. In the deep one questions about very specific data may be asked by the system during the process of instantiation of the causal network, but such questions are isolated, so that there is not a problem of massive data acquisition.

We have noted before that the surface (heuristic) level of CHECK is formed by a hierarchy of frames; in the following we will sketch the organisation of each frame (for more details see [18][19]).
Among the different groups of slots in a frame, two are important to make the following description clearer:
- prototypical description slots
- control slots
The prototypical description is divided into two parts: MAJOR FINDINGS and SPECIFIC FINDINGS. The first contains the list of the findings that are necessary to give a first (general) characterisation of the diagnostic hypothesis, the second the list of more specific findings that refine the characterisation of a particular diagnostic hypothesis.
Different types of control slots are present in a frame, among them TRIGGERS, which are activation rules (i.e. a frame is considered only if these rules can be satisfied by the data relative to the case under examination) and CONFIRMATION RULES, which are rules involving very specific data (e.g. laboratory examinations) used to refine diagnostic hypotheses (in particular they are used to differentiate between hypotheses (see [18][19] for more details) must be remembered.

The data acquisition interface of CHECK is based on the use of graphical menus, which have to be filled by the user with proper data (findings and their linguistic values, see Fig. 1). We have recognised, following the scheme of section 2, two distinct phases of data acquisition: an initial one in which the system is passive and one in which the system actively requests data.
The passive phase starts just at the beginning of a consultation, i.e. is the first operation performed by the heuristic module. A graphical menu, which has to be filled by the user with initial data, is displayed by the system. For avoiding the introduction of unknown findings the menu contains a list of items (findings) which have to be selected by the user. In particular only very general findings are inserted in such menu,

that is only the findings needed to verify the triggering conditions of the higher level diagnostic hypotheses (classes of diseases). When the user selects a finding in the menu a second pop-up menu is displayed by the system containing the possible linguistic characterisations of the finding itself, which have to be selected by the user. It is worth noting that default (normal) values are automatically associated by the system to all the findings which have not been selected by the user from the main menu.

Such initial data are sufficient to evocate (trigger) initial (coarse) diagnostic hypotheses (frames). During the process of instantiation of the frames corresponding to such hypotheses more specific data (those contained in the MAJOR FINDING slots and then, if the evidence degree of the match beteen MAJOR FINDINGS and actual data is strong enough, those in the SPECIFIC FINDINGS slot) are requested to the user via graphical menus. It is worth noting that only clinical data and very simple laboratory tests results are asked in this phase, moreover only data relative to relevant hypotheses (i.e. those evocated by initial data) are requested. During the refinement and specialisation processes of the initial hypotheses other very specific data (e.g. about laboratory examinations) may be asked by the system in the same way. During all the data acquisition process the user can ask information (explanations) about data, which is displayed on a window on the screen. Moreover the system displays, for the whole consultation, the clinical record of the patient in a different window (see Fig. 1).

It is worth noting that this scheme for data acquisition matches the ideal scenario described in section 2, i.e. the natural process adopted by the physician visiting a patient.

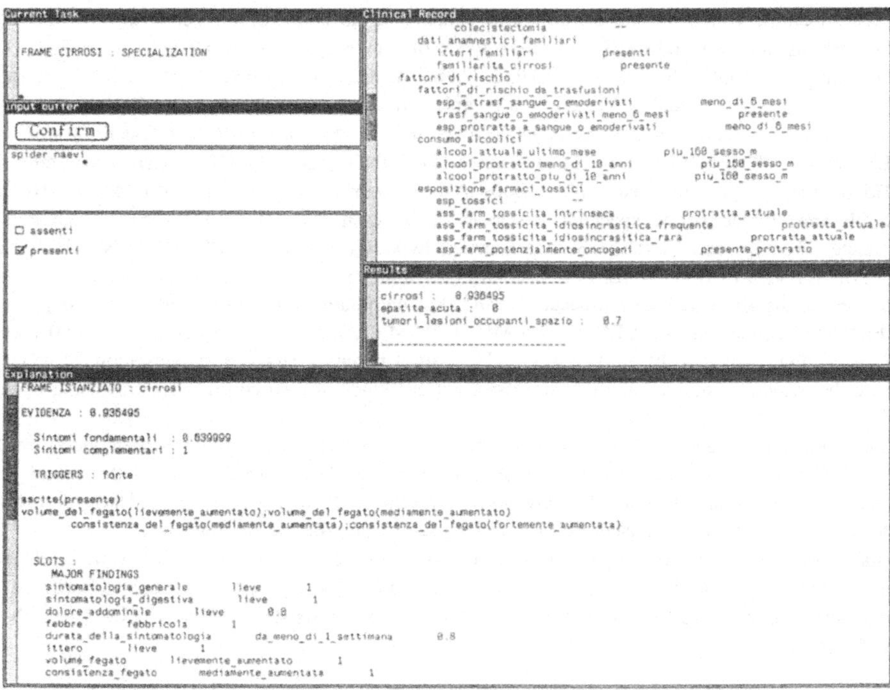

4.2 Explanation

We have noticed above that the use of deep knowledge is mandatory for the system could generate precise explanations about its advices, in term which are familiar to a physician. This does not mean that in the heuristic level no explanation component is needed, in fact we have provided simple mechanisms at this level too. In particular the system provides menus for the display of instantiated frames (i.e. diagnostic hypotheses and data which allowed their activation), in a specific explanation window (see Fig. 1) and to get information about data.

Nevertheless the major part of the explanation generation in CHECK is charged to the deep level. The knowledge representation formalism we have designed for the deep level of CHECK is based upon the use of causal networks.

Five types of nodes are defined in the causal network of CHECK: HYPOTHESES, STATES, INITIAL CAUSES, FINDINGS and ACTIONS. The nodes of type HYPOTHESIS represent the diagnostic hypotheses in the system and correspond to the frames in the surface level. STATE nodes represent possible conditions of the modeled system (e.g. a pathophysiological condition in a physiological model). Each STATE is characterised by a set of state variables, which are used to describe the state itself. The characterisations we use are qualitative ones, that is each variable can assume only one of a set of possible linguistic values. These values are set during a particular consultation using qualitative reasoning techniques. INITIAL CAUSES nodes represent the possible deep causes of the arising of a failure (disease in a medical domain). The FINDING nodes are the manifestations of STATES and HYPOTHESES. ACTION nodes are used to model events in the system: two states can be logically connected by a cause-effect relation just when an action (event) causes the transition.

Let us consider now which are the possible arcs connecting the nodes of the network, i.e. which are the possible relations between nodes. Six different types of arcs are defined in CHECK: CAUSAL, LOOP, HAM, FORM_OF, SUGGEST and CONCLUDE. CAUSAL arcs model cause-effect relationships: each one of them connects two states and is formed by a complex structure which involve an ACTION node and two sub-arcs: a STATE-ACTION and a ACTION-STATE arc. HAM (Has As a Manifestation) arcs connect STATES (or HYPOTHESES) to their manifestations (FINDING nodes). LOOP arcs are structurally similar to CAUSAL one, but they are used to close loops in the network. The other types of nodes in CHECK are: FORM_OF arcs, which define specialisation relations between HYPOTHESES nodes, SUGGEST arcs, which relate STATES and FINDINGS to associated hypotheses and CONCLUDE arcs, which causally relate set of STATES to HYPOTHESES. Each arc (transition) in the network can be labeled as a MAY or a MUST arc, to discriminate between necessary and possible relations. Let us consider for example a HAM arc: a FINDING connected to a STATE by a MUST HAM arc represents a necessary (definite) manifestation of the STATE, whilst if the arc is a MAY HAM arc it is only a possible manifestation. Moreover conditions can be associated to each arc, which impose various kinds of restrictions to the traversing of the arc.

The explanation for a proposed diagnosis is generated after the hypothesis itself has been deeply confirmed on the causal model. Such explanation consists mainly in the display of the part of the causal network which has been instantiated to confirm a diagnostic hypothesis. In this way a causal description, from the ultimate causes to the final manifestations of a disease can be provided to the user.

The instantiated part of a network is displayed by the system on a special graphical window; for the sake of clearness information about the instantiation of states (and other nodes) is provided separately (in a different window), when the state itself is selected by clicking the mouse on its shape in the window where the net is displayed (see Fig. 2).

A major problem in the visualisation is connected with the dimensions of the network (not only is this is a problem for explanation, but, more generally for the whole reasoning process on causal models, see [13] [17] for a discussion), which cannot be displayed on a screen. To solve this problem we have decided to design causal networks at different levels of abstraction: at the first level only few nodes are present (INITIAL CAUSES, HYPOTHESES and very general states). Such high level network (or the part of it which has been instantiated) is displayed first to give an immediate and clear explanation about the pathophysiological state of the patient. General states of the higher level network can then be elaborated, that is they can be expanded in a more precise and complex network. Such network can be displayed by selecting the appropriate high level node with the mouse.

Moreover when we display part of a complex network, which cannot be visualised in a screen, we give, in a separate window, the global form of the network in a reduced scale, in which the displayed part is highlighted.

The system provides also the possibility of inspecting the general form of the causal network, independently on the data of a particular patient, to get precise explanations about the domain of expertise of the systems [17]. Such forms of explanation are very useful, maily for tutorial purposes.

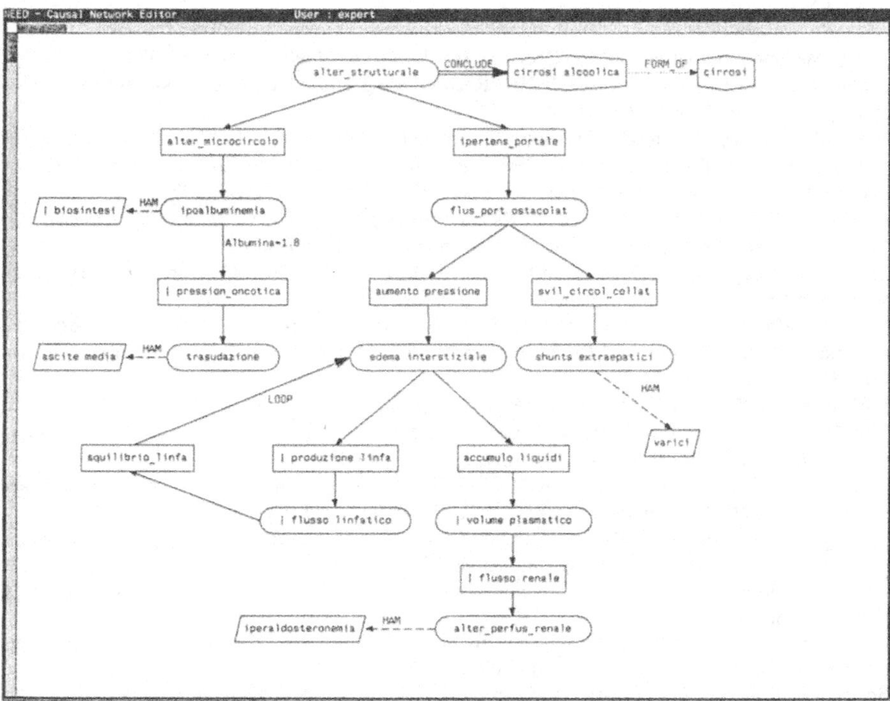

Note that elliptic boxes represent states (e.g. "ipoalbuminemia" or "reduced plasma albumin concentration"), rectangle boxes represent actions (e.g. "riduzione pressione oncotica", or "oncotic pressure reduction"), exagonal boxes represent diagnostic hypotheses (e.g. "cirrosi" or "cirrhosis") and rhomboidal boxes represent findings (e.g. "ascite media" or "mild ascites"). The type of each arc is displayed near the arc itself, execpt for merely causal arcs.

Figure 2

5. CONCLUSIONS

In this paper we have presented some considerations about the construction of a man-machine interface for a medical diagnostic expert system. These considerations have led to the definition of an ideal scenario for such an interface. We have then described the graphical interface we have designed for CHECK, showing that it is a practical implementation of such an ideal scheme.

It is worth noting that we have designed also flexible interfaces for the knowledge engineer, in particular we have designed a graphical editor for the design of causal networks, which is able to perform both syntactic and semantic checks on the consistency of a deep model and which can generate a linguistic description of the network, to be used by the consultation module (for more details see [17]).

CHECK is implemented in Prolog, using a graphical extension of C-Prolog we have implemented on SUN workstations. The graphical components (interfaces) are implemented partially in Prolog and partially in 'C' language (the lower level procedures).

REFERENCES

[1] Fox, J.: "Medical Computing and the User"; in *Int. J. of Man-Machine Studies, 1977.*

[2] Clancey, W.J., Letsinger, R.: "NEOMYCIN: Reconfiguring a rule-based expert system for application to teaching"; in: *Proc 7th IJCAI* 1981, 829-836.

[3] Fieschi, M., Joubert, M., Fieschi, D., Botti, G., Michel, C., Proudhon, H.: "The Sphinx Project"; in I.De Lotto and M.Stefanelli(eds.) : *Artificial Intelligence in Medicine,* North-Holland, 1985, 83-93.

[4] Shortliffe, E.H.: "*Computer based Medical Consultation: MYCIN"; Elsevier 1976.*

[5] Buchanan, B.G., Shortliffe, E.H. (eds): "*Rule Based Expert Systems: The MYCIN Experiments at the Stanford Heuristic* Programming Project; Addison Wesley 84; cap. 17.

[6] Carbonell, J., Boggs, W.M., Mauldin, M.L., Anick, P.G.: "The XCALIBUR project: a natural language interface to Expert Systems"; in *Proc 8th IJCAI* 1983, 653-656.

[7] Fagan, L., Differding, J., Langlotz, C., Tu, S.: "Knowledge Acquisition and Strategic Therapy Planning for cancer Clinical Trials" in I.De Lotto and M.Stefanelli(eds.) : *Artificial Intelligence in Medicine,* North-Holland, 1985, 75-81.

[8] Shortliffe, E.H., Scott, A.C., Bischoff, M.B., Campbell, A.B., van Melle, W., Jacobs, C.D.: "ONCO-CIN: An expert system for oncology protocol management". In: *Proc 7th Annual Symposium on Computer Applications in Medical Care.* Baltimore, 1983: 149-152.

[9] Richer, M.H., Clancey, W.J.: "GUIDON-WATCH: A Graphic Interface for Viewing a Knowledge-Based System"; *IEEE Computer Graphics,* 5,11 November 1985, 51-64.

[10] Clancey, W.J.: "From GUIDON to NEOMYCIN and HERACLES in Twenty Short Lessons: ORN Final Report 79-85"; in *The AI Magazine,* August 1986, 40-60.

[11] Trappl, R., Horn, W.: "Making Interaction with a Medical Expert System easy"; in *Proc MEDINFO 83,* North Holland 1983.

[12] Fox, J., Frost, D.: "Artificial Intelligence in Primary Care"; in I.De Lotto and M.Stefanelli(eds.) : *Artificial Intelligence in Medicine,* North-Holland, 1985, 137-154.

[13] Molino, G., Cravetto, C., Torasso, P., Console, L.: "CHECK: A Diagnostic Expert System Combining Heuristic and Causal Knowledge"; to appear on *Biomedical Measurement, Informatics and Control,* 1987.

[14] Clancey ,W.J.: "The Epistemology of a Rules Based Expert System: a Framework for Explanation"; in *Artificial Intelligence,* 20 (3), 1983: 215-251.

[15] Wallis, J.W., Shortliffe, E.H.: "Customized Explanations Using Causal Knowledge"; in Buchanan, B.G., Shortliffe, E.H. (eds) "*Rule Based Expert Systems: The MYCIN Experiments at the Stanford Heuristic* Programming Project; Addison Wesley 84; cap. 20.

[16] Steels L.: "Second Generation Expert Systems". *Fifth Generation Computing Systems* 1985; 1: 213-221.

[17] Torasso, P., Console, L.: "Causal Reasoning in Diagnostic Expert Systems"; *Proc. SPIE Conf. on Applications of Artificial Intelligence V,* May 1987, Orlando.

[18] Cravetto, C., Lesmo, L., Molino, G., Torasso, P.: "LITO2: A Frame Based Expert System for Medical Diagnosis in Hepatology"; in I.De Lotto and M.Stefanelli(eds.) : *Artificial Intelligence in Medicine,* North-Holland, 1985, 107-119.

[19] Cravetto, C., Lesmo, L., Massa Rolandino, R., Molino, G., Torasso, P.: "An expert system for liver disease diagnosis (LITO2)". In: *Proc 9th Annual Symposium on Computer Applications in Medical Care.* Baltimore, 1985: 330-334.

The Oxford System of Medicine:
A prototype information system for primary care

John Fox, Andrzej Glowinski, Michael O'Neil

Imperial Cancer Research Fund Laboratories
Lincoln's Inn Fields
London WC2A 3PX

SUMMARY

The Oxford System of Medicine (OSM) is an experimental knowledge based system designed to satisfy requirements of information provision and decision support in general practice. A system of this type must address special problems caused by the wide medical scope of general practice; the variety of decisions taken by general practitioners, and the acute user-interface demands in this clinical setting. The aims of the OSM project, and the design and operation of the prototype are outlined. Three central aspects of the system are discussed: the user interface, organisation of the knowledge base, and the generalised approach to decision making developed during the project.

HISTORY OF THE PROJECT

Oxford University Press is a major medical publisher, also heavily committed to electronic publishing. During 1985 the Press carried out a study into whether a new kind of medical information service, to be called the Oxford System of Medicine, might be developed for use in the 1990s. The OSM would be a computerised information system designed to provide information on a wide range of clinical topics, and able to assist GPs in making a variety of medical decisions. It was viewed as a clinical tool for doctors having to keep pace with the large and growing body of modern medical knowledge. It was clear that development of the OSM would require considerable resources, but it was uncertain whether a practical design was theoretically or technically viable.

The Biomedical Computing Unit of the Imperial Cancer Research Fund has an established program of research into medical decision making, and a special interest in computer systems for general practice. Effective screening, early detection, accurate diagnosis and management of cancer, can all benefit from computer technology, though only as an integrated part of general medical practice. The ICRF and OUP agreed to collaborate on a short term technical study during 1986. Earlier work on interactive systems for clinical decision support (Fox, Barber and Bardhan, 1979;1980) on the design of the user interface and the clinical impact of such systems (Fitter and Cruickshank 1983; Barber and Fox, 1981), and the establishment of a design for a knowledge based general practice information system (Fox and Rector, 1982; Fox and Frost, 1985) provided the starting point for the study.

THE PROTOTYPE CONCEPT

The concept of the OSM is illustrated in figure 1. A small personal computer on the doctor's desk (independent or linked to other machines in the practice) is immediately to hand during a patient consultation.

Figure 1 The OSM is seen as an integral part of the general practice consultation. The computer and mouse are immediately to hand during the appointment.

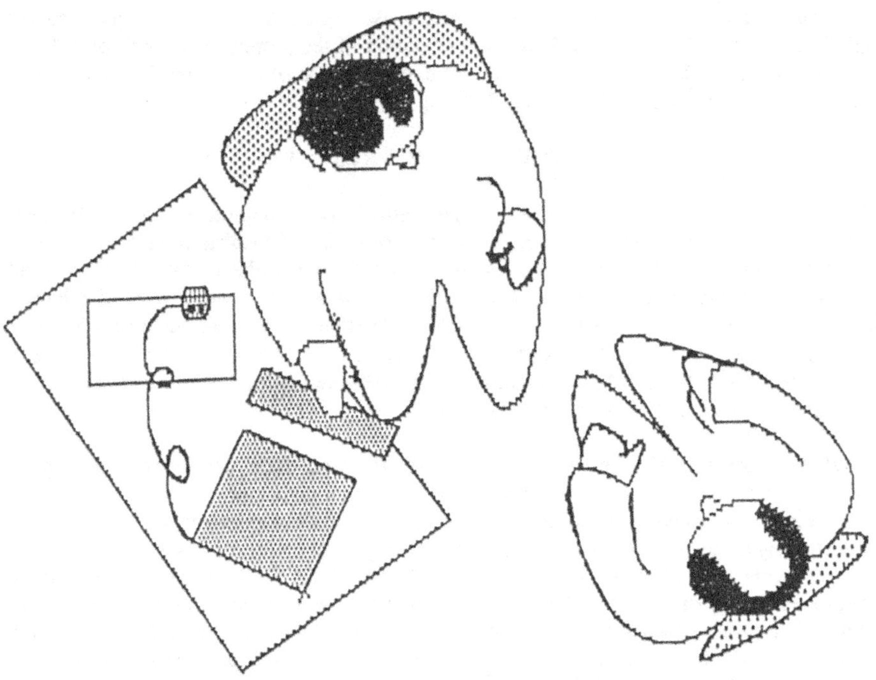

The OSM must provide information in a variety of ways. Frequently, doctors may only require a quick, convenient way of getting access to information - some detail about a patient, drug, disease, investigation or local service. Sometimes, however, the doctor may find it useful to have assistance in making a specific diagnostic or management decision. In this circumstance the computer must be able to automatically locate any information from both the medical knowledge and patient record, however obscure, which could be relevant to the decision. Occasionally the doctor may even value a "second opinion": What is the most plausible diagnosis? What would be the best treatment or investigation? How might the treatment or investigative procedure be planned?

AIMS OF THE PILOT PROJECT

The OSM pilot project explored the practical and theoretical possibility of developing a system with these characteristics. The six month study started with three specific aims.

1. To assess whether an OSM is technically feasible and suitable for development, or whether it is still a distant goal requiring further research.

2. To explore the facilities that might be needed for an OSM to have practical value to general practitioners and to develop a prototype incorporating some of these facilities.

3. To consider whether the Oxford System of Medicine could be designed in such a way that a routine procedure could be established for electronic publication of medical knowledge to high medical and technical standards.

THE PROTOTYPE DESIGN

Figure 2 illustrates the user's view of the prototype. This layout has three main areas. At the top are four panels, labelled "information", "assistance", "synopsis" and "explanation". These provide menu-based access to four groups of facilities: access to information about patients, drugs, diseases, investigations and treatments; an "assistant" for decision making; facilities for summarising and reviewing the case; explanations of questions, conclusions, the decision making method being used, and so on. Other facilities would be needed in a mature system. The large centre panel is used for presenting information to the user, and for entering patient data using a mouse or other pointing device. The small bottom panel (which is optional) allows direct access to Props 2, the development package which was used for the implementation of the OSM prototype (Frost et al 1986).

Figure 3 illustrates the use of the OSM prototype with a simple sequence of screens produced while looking up prescribing information on the drug cimetidine.

Figure 4 illustrates the use of the assistant, in this case to prescribe morphine.

Figure 5 illustrates a slightly more complex use of the assistant in the classification (diagnosis) of joint pain.

THE USER INTERFACE

Ease of use and understanding are vital to the acceptability of computer systems, perhaps nowhere more so than in medicine. One of the contributions of early medical expert systems (notably Mycin; Buchanan and Shortliffe, 1984) was to emphasise the importance of good user interface design. In general practice time pressures and the unpredictable nature of the work make the design of the OSM's user interface particularly critical. To achieve acceptability several obvious requirements must be

Figure 2

The user's view of the OSM: (a) Information retrieval from the knowledge base. (b) Assistance with making decisions (c) Reviewing and summarising the case (d) Various methods of explanation

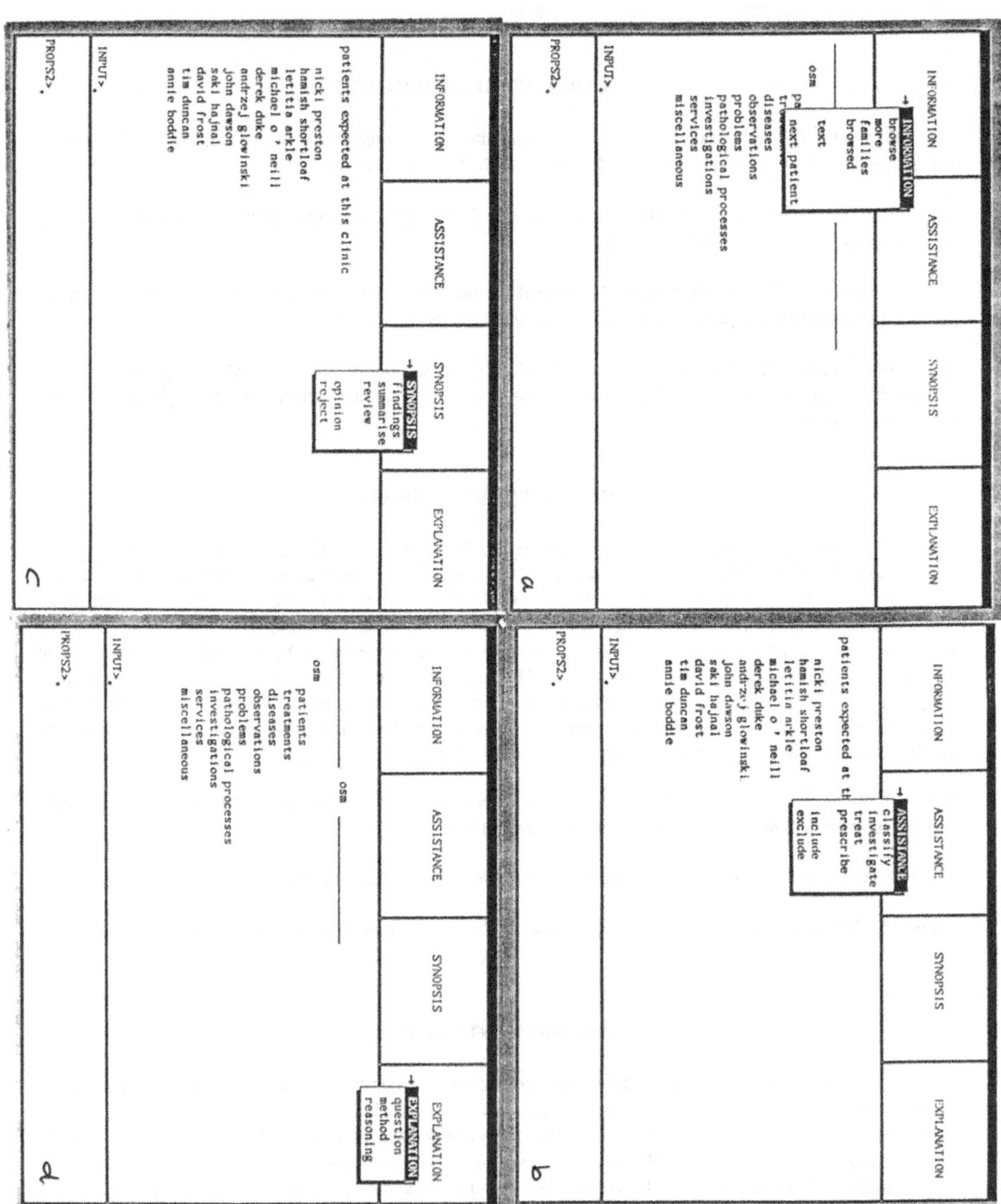

Figure 3

Browsing information the knowledge base. (a) Subdivisions of the topic being looked up are presented. Gastrointestinal drugs have been selected as the next topic using the mouse. (b) A subdivision of gastrointestinal drugs - cimetidine is being chosen. (c) As well as subdivisions (tagamet) various items of information about cimetidine are available. (d) Contraindications of cimetidine have been selected.

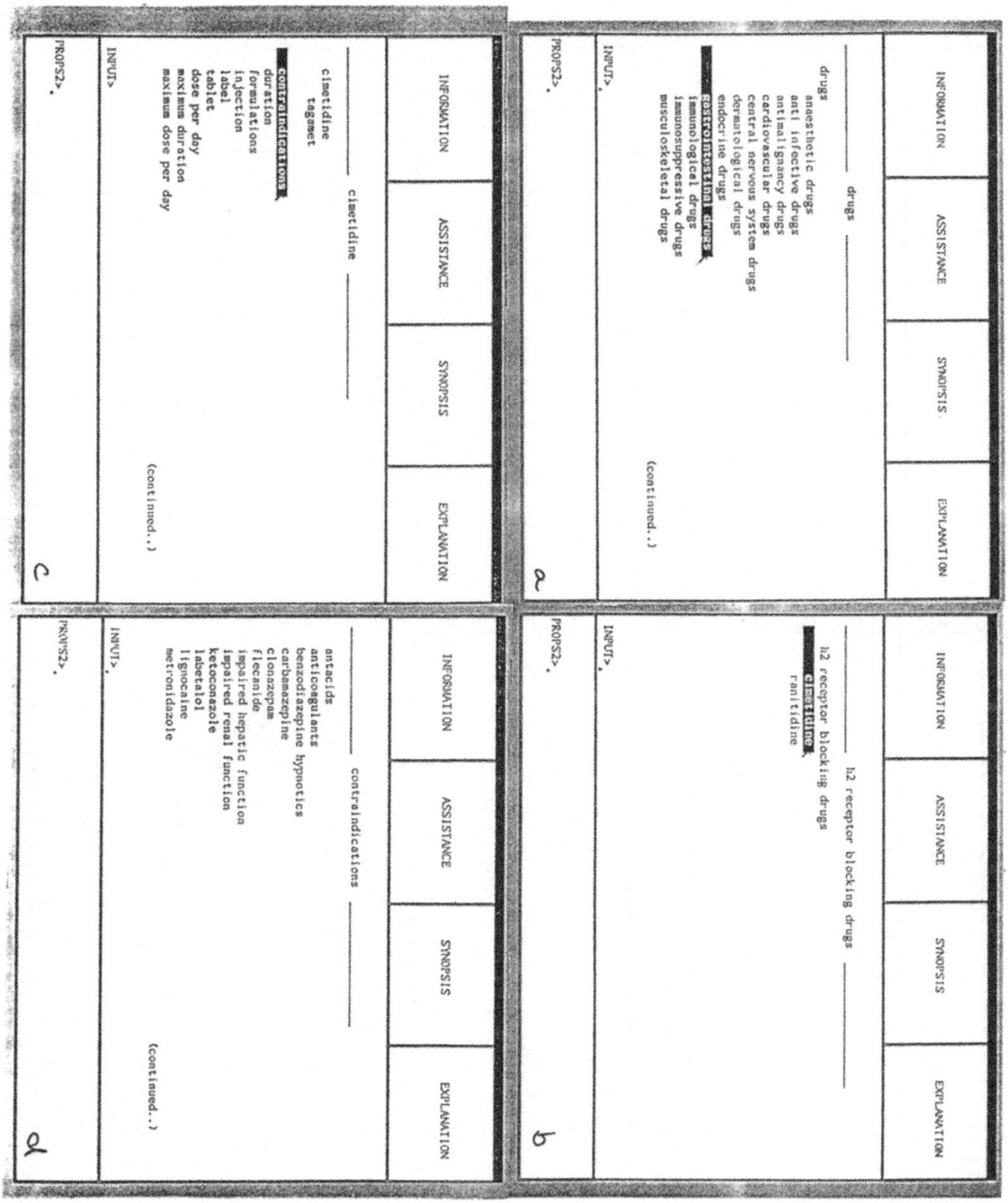

Figure 4

Prescribing morphine. (a) Selecting the 'prescribe' option from the Assistance menu. (b) Indicating
which drug to prescribe by selecting it - typing the name can save time if the drug does not appear on
the screen. (c) Items relevant to the prescription of the drug are presented - selecting any of these
affects the final prescription. (d) The prescription - note that the prescription must be hand written as
morphine is a controlled drug. The user is at liberty to change any of the chosen values, the change being checked against acceptable limits and a warning given if these are not met.

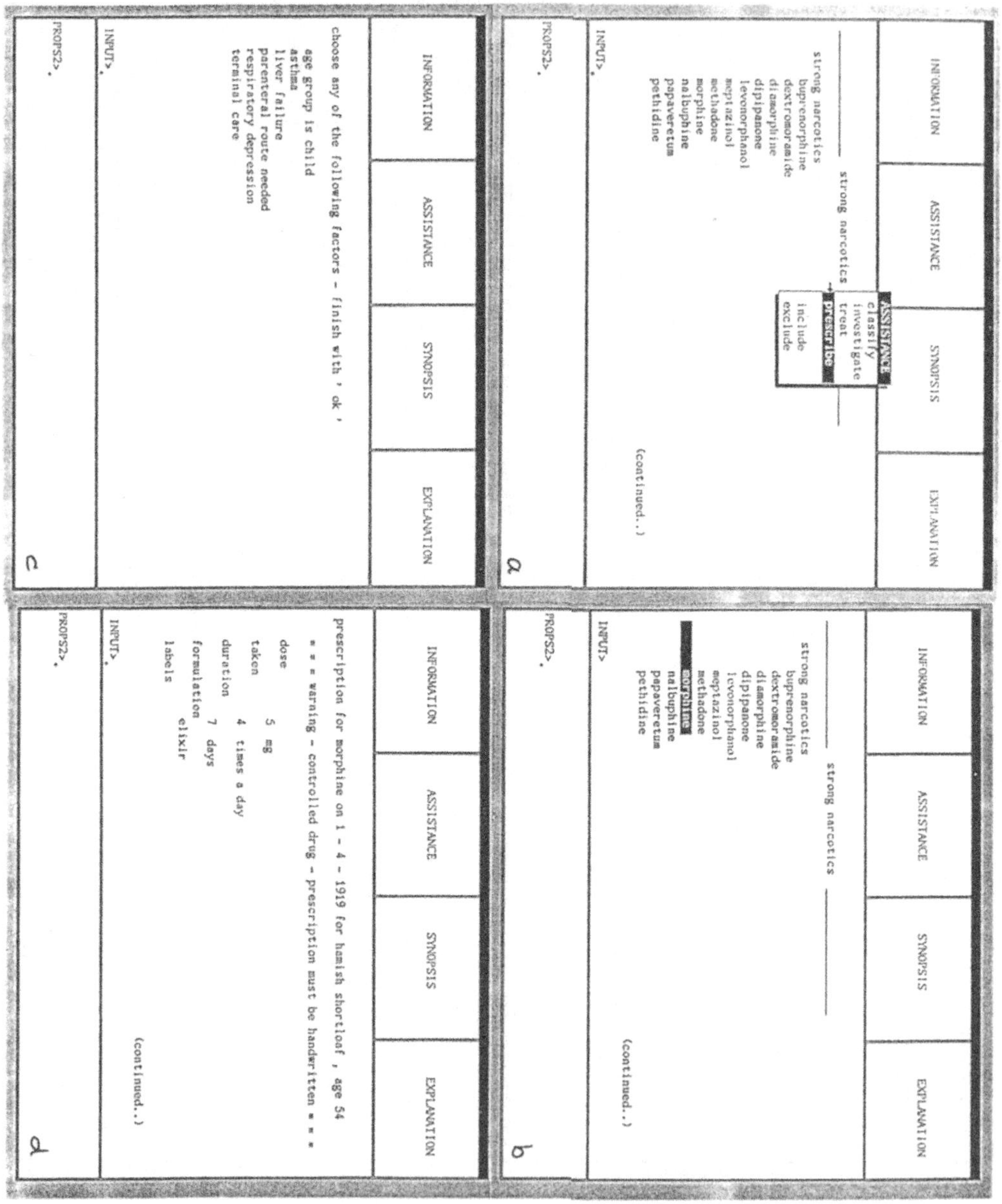

(a)

INFORMATION | ASSISTANCE | SYNOPSIS | EXPLANATION

```
________ strong narcotics ________
         strong narcotics
           buprenorphine
           dextromoramide      ASSISTANCE
           diamorphine         classify
           dipipanone          investigate
           [levonorphanol      treat
           meptazinol          prescribe
           methadone           include
           morphine            exclude
           nalbuphine
           papaveretum
           pethidine
```

(continued..)

INPUT>.
PROPS2>.

(b)

INFORMATION | ASSISTANCE | SYNOPSIS | EXPLANATION

```
________ strong narcotics ________
         strong narcotics
           buprenorphine
           dextromoramide
           diamorphine
           dipipanone
           [levonorphanol
           meptazinol
           methadone
           morphine
           nalbuphine
           papaveretum
           pethidine
```

(continued..)

INPUT>.
PROPS2>.

(c)

INFORMATION | ASSISTANCE | SYNOPSIS | EXPLANATION

```
choose any of the following factors - finish with ' ok '
age group is child
asthma
liver failure
parenteral route needed
respiratory depression
terminal care
```

INPUT>.
PROPS2>.

(d)

INFORMATION | ASSISTANCE | SYNOPSIS | EXPLANATION

```
prescription for morphine on 1 - 4 - 1919 for hamish shortloaf , age 54
= = = warning - controlled drug - prescription must be handwritten = = =

dose          5   mg
taken         4   times a day
duration      7   days
formulation   elixir
labels
```

(continued..)

INPUT>.
PROPS2>.

Figure 5

A sequence illustrating some of the stages of classification (differential diagnosis). (a) Selecting the problem the mouse while browsing joint observations (b) Information relevant to the task is presented in the form of menus. As more information is added the range of items expands to reflect the current hypotheses. (c) The reason for any item appearing on the screen can be requested using the set of options in the 'Explanation' menu. (d) A summary of conclusions can be requested at any stage - this is an example of a report part way through the classification of joint pain.

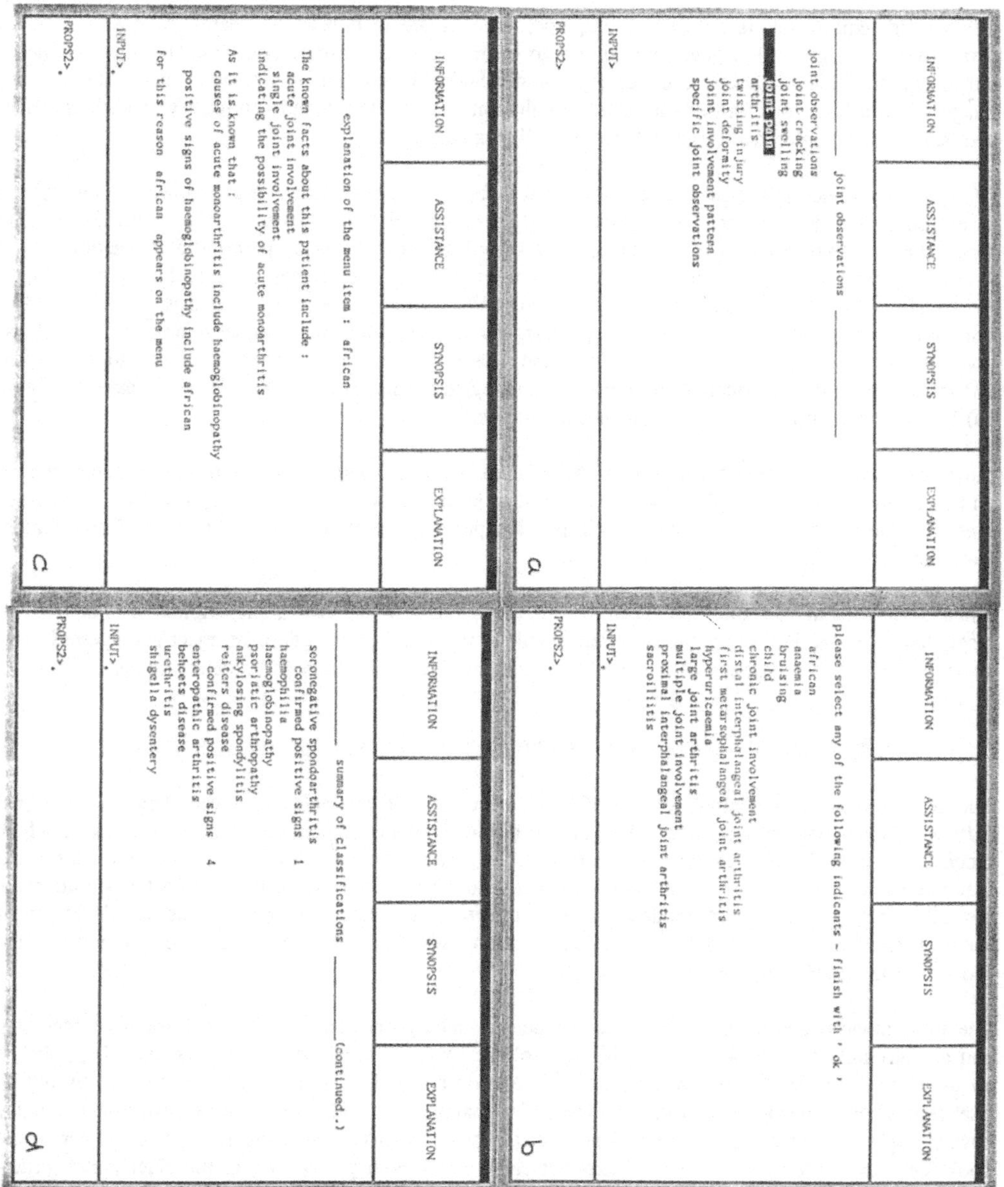

satisfied.

The first requirement is simplicity of operation. The OSM prototype makes use of standard WIMPS tools; the mouse can be used to operate control menus, issue commands or enter items of data by pointing at a word or phrase on the screen, etc. Pressing a button mounted in the mouse confirms the selection.

The second requirement is conceptual simplicity. The facade in figures 2-5 seems to be successful in that it is easy to grasp and physically simple to operate with few errors (see note 1). Many enhancements are possible, including graphics (eg for displaying blood pressure graphs), static and moving images from a laser disk (such as a rash or an abnormal gait). The added complexity of such enhancements, however, could have disadvantages as well as advantages.

Of equal importance is the subtler requirement that the system be responsive to changing requirements, as judged by the practitioner. The system should be controlled by the user, not vice versa. Props 2, in which the OSM prototype is implemented, is designed to support this by permitting any action by the user at any time without intricate command protocols. The OSM prototype extends this by providing a mechanism which allows interruption of any activity (information retrieval, diagnosis, treatment selection etc) at any point by starting another process. For instance, during diagnosis summaries or explanations of the current conclusions can be obtained, possible treatments considered or any topic "looked up" (by invoking the information browser and pointing to the required topic with the mouse - see figure 2a) before resuming the interrupted diagnostic process.

Since the OSM is intended for use in the clinical setting, where information requirements can be unpredictable and change rapidly, we believe that at least this degree of flexibility is critical. The OSM must provide a range of information facilities; the prototype provides these but in an elective rather than dictatorial style.

Finally, the system can offer an 'opinion' on a diagnosis or suggest a prescription if, but only if, instructed to do so. Even here the user can modify a suggested prescription, or reject a suggested diagnosis, or instruct the computer to include or exclude specific possibilities in its deliberations.

PUBLICATION AND ORGANISATION OF THE OSM KNOWLEDGE BASE

The development of an Oxford System of Medicine could not be treated as a very large conventional software engineering project, like a banking or airline booking system say, some examples of which incorporate upwards of a million lines of computer code. The complexity and cost of creating and maintaining a conventional system which covered enough medicine to be useful to a GP would make the OSM quite impractical. A fundamental requirement of the design, therefore, was that development should take place under the direct editorship of authoritative medical professionals, employing adapted forms of established editorial methods.

The most obvious approach would be to pursue an analogy between the Oxford System of Medicine and a multi-author textbook, such as the two volume Oxford Textbook of Medicine (OTM) published by the Oxford University Press. Chapters in this general medical reference work are prepared by individual specialists, under loose editorial guidance. By analogy in the electronic version we might imagine developing a number of loosely related specialist expert systems, each comparable to a chapter in the book. One specialist would deal with terminal care; others with prescribing in the elderly and gastro-intestinal diseases, and so forth. Each system would be largely independent, though presumably with some overall standards concerning the user interface and general approach.

This method soon shows itself to be inappropriate because it fails to address the need to build and maintain to a high, consistent standard a structure of the size and scope of an OSM. It would permit each expert system to embody different medical assumptions and strategies. The knowledge required in different areas could overlap, yet the knowledge organisation used in one module might be incompatible with that in another, leading to unpredictable (and potentially undetectable) gaps and interactions in the knowledge base. Development would be costly, performance poor and different medical areas inconsistent in their operation and facilities.

We therefore adopted a different approach which was intended to promote a consistent structure across medical subfields, rather than self-contained specialist systems. The knowledge in the prototype design is organised as sets of modules arranged in layers as illustrated in figure 6.

Figure 6

A model of the structure of the (first) OSM prototype. Modules are arranged into increasingly specific layers of knowledge with increasing distance from the core.

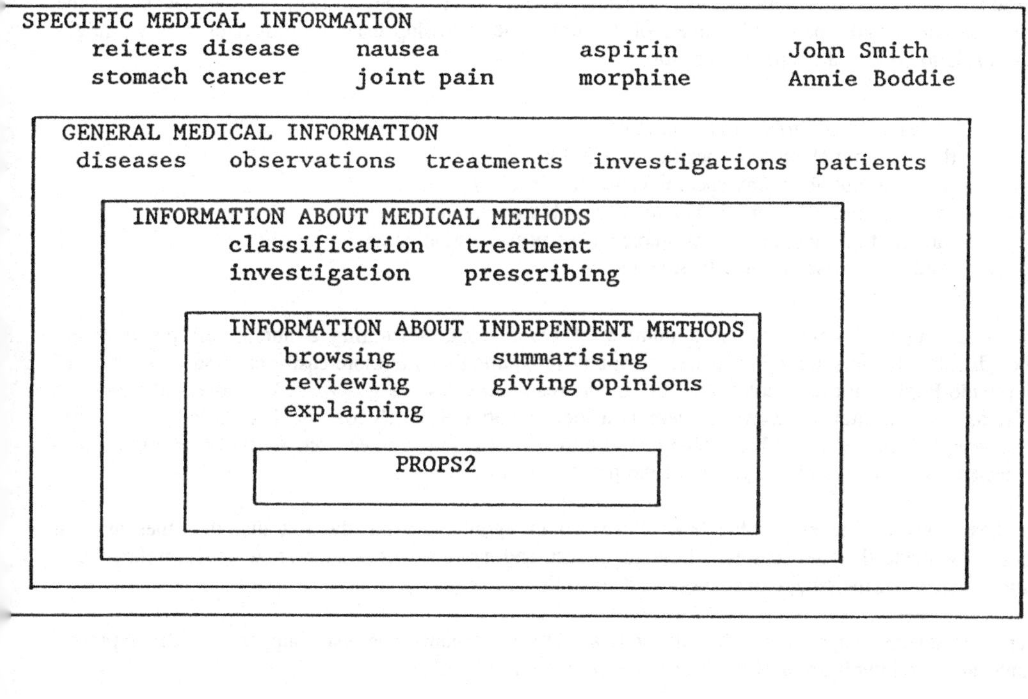

```
SPECIFIC MEDICAL INFORMATION
      reiters disease    nausea        aspirin       John Smith
      stomach cancer     joint pain    morphine      Annie Boddie

    GENERAL MEDICAL INFORMATION
      diseases    observations   treatments   investigations   patients

        INFORMATION ABOUT MEDICAL METHODS
              classification    treatment
              investigation     prescribing

            INFORMATION ABOUT INDEPENDENT METHODS
                browsing          summarising
                reviewing         giving opinions
                explaining

                        PROPS2
```

Each module contains information about a topic, written and stored in the computer in a notation that we call "pseudo-English" (Frost et al, 1986). The outermost layer in figure 6 contains sets of medical facts about particular patients, drugs, diseases etc.

    problems of John Smith include gout
    side effects of narcotics include constipation
    positive signs of peptic ulcer include epigastric tenderness

The adjoining layer of modules contains information relating topics to each other. A straightforward example is the "drug hierarchy". Different classes of drugs, described in publications such as the British National Formulary, are represented in the OSM by pseudo-English facts::

    kinds of drugs include cardiovascular drugs
        kinds of cardiovascular drugs include diuretics
    kinds of drugs include gastrointestinal drugs
        kinds of gastrointestinal drugs include ulcer healing drugs
            instances of ulcer healing drugs include cimetidine ... etc

The next layer is rather different because the modules embody knowledge about how to make medical decisions - diagnosis, treatment etc. Consequently, as well as facts, they contain rules which specify the steps required to make particular classes of decision. The following example illustrates how we might write a rule to control a treatment decision:

    if treatment of Problem is required
    then find out about any specific form of Problem
    and find out about any specific causes of Problem
    and find out about any indicants of possible treatments
    and find out about any contraindicants of possible treatments
    and summarise possible treatments

This is a simplified composite of a number of OSM rules, illustrating a number of points. First, although rules like this are equivalent to computer programs they are more easily understood because of the pseudo-English form. Second, this and other treatment rules are generalised to cover all treatment decisions. The capitalised word Problem is a logical variable which may bind to "joint pain", "anxiety", "weight loss", or any other problem whose treatment requires a decision. A consistent approach to the decision irrespective of the details of the problem results.

The final layer of the knowledge base illustrated in figure 6 is like the last that the rules are not specifically medical. They describe how to collect and present information, how to formulate summaries, reviews of the data, explanations and so forth.

There are several ways in which this modular, layered organisation may support the development, maintenance and publication of medical knowledge on a large scale.

(1) Information about medical topics and decision making methods is in pseudo-English, so is relatively intelligible to editors with more knowledge of medicine than of computer systems (but see note 2).

(2) The great bulk of the medical knowledge in the system (perhaps a million or more facts in a first edition) is represented as a database, not as a program. The programs required (generalised medical and domain independent methods) are small in comparison, perhaps 200-300 rules. Database

maintenance is a well established topic, and parts of the process could be automated; maintaining programs is far more difficult.

(3) A clear organisation of the information results from the modularity, making the editorial process simpler.

(4) Representing medical decision making as a generalised, intelligible procedure allows critical analysis by professionals inside and outside the design team.

A GENERALISED APPROACH TO DECISION SUPPORT

Knowledge based systems emphasise the role of qualitative reasoning in formulating their decisions. The OSM is no exception. However the decision making mechanisms of most medical expert systems (most famously in MYCIN, Casnet and Internist) have also made use of quantitative methods for calculating disease probabilities, the expected value of treatments and so on. Although combining quantitative methods with qualitative logical techniques has not been excluded from a future Oxford System of Medicine, no particular emphasis has been placed on this in the prototype.

There are a number of reasons for this apparent omission. Firstly, the available techniques for combining logic and probability are, in our view, ad hoc and not based on sound, well understood principles (Fox, 1985). Secondly, effective use of mathematical methods depends on the availability of well estimated statistics, covering the incidence of symptoms and disease, the reliability and value of investigations and so on. These statistics are only available in a small proportion of the areas which concern the general practitioner. Thirdly, we place value on the intelligibility of the OSM to the medical professional whereas mathematics may detract from this and turn the support system into a mysterious black box. Quantitative methods offer more precisely calculated decisions, but many practitioners are not convinced that increased precision brings enough advantages to risk losing intelligibility.

One of the great strengths of mathematical decision making, however, is that its principles are well understood, while those that underly knowledge based systems are relatively novel. We therefore accept a responsibility to develop a framework for decision support in the OSM whose principles are as well motivated as classical mathematical methods, or are at least equally well described. Satisfying this requirement is currently a major focus of our continuing work on the OSM; we can give a provisional outline here, though a detailed presentation must await a more comprehensive technical paper.

In essence we are trying to state the body of principle that underlies various forms of medical decision. In the OSM decision making processes are specified as general, meta-level rules (Fox, 1984). These are general in the sense that they are independent of the problem which is being considered. In fact even more general procedures may underly all of the decisions which the OSM will support, a possibility we return to later. (Note that these processes are user and data-driven and effectively operate asynchronously and largely in parallel, rather than as preprogrammed, sequential steps.)

In the current prototype decisions are viewed as choice procedures. Suppose the OSM is required to assist in selecting a treatment for a patient with nausea. Given only this information about the patient some possible causes of the nausea will be apparent (ie from facts like "possible causes of nausea include biliary disease") and some obvious possible treatments ("treatments of nausea include antiemetics"). As more information about the kind of nausea is obtained (eg acute or periodic) the set of possible causes propagates and consequently the set of possible treatments may increase or decrease.

For each possible choice of treatment there may be specific supporting or weakening factors, such as positive or negative evidence, indications or contraindications (the knowledge base contains facts of the form "contraindications of antiemetics include pregnancy"); those that are confirmed by the GP represent "pros" and "cons" of the possible treatments.

Pros and cons refer to individual choices; logical interactions between choices must be taken into account as well. For example antiemetics may be indicated for a patient, but a more specific indication for metoclopramide may also be present, so the decision procedure should recognise that one may subsume the other ("kinds of antiemetics include metoclopramide") and disregard the more general one. Many relationships can be exploited to reduce the number of possibilities, including relative cost, acceptability to the patient, and so on.

Sometimes a single preferred choice (or combination of choices) remains. If a recommendation is requested this is the one to be offered. Often, however, more than one possibility will remain and an overall quantitative assessment of desirability will be necessary. The last component of the decision procedure is an aggregation mechanism for combining the pros and cons in order to make a choice (note 3). In the present system these pros and cons are not weighted quantitatively, though in principle they might be (see remarks above).

Summarising the main components of the decision procedure:

1. Generate the possible options (diagnoses, investigations, treatments) through propagation over causal, taxonomic and other structures.

2. Identify the pros and cons for each option as specified in meta-knowledge of the different kinds of decision.

3. Identify interactions between the possible choices according to criteria recorded in the knowledge base and prune the set accordingly.

4. Establish the balance of preference when a single choice must be made

This approach to decision making, as explicit representation of general methods is to be contrasted with the implicit representation of decision strategy in large numbers of specialised domain rules. Domain knowledge is stored separately in the fact base. Among the advantages of this approach are the relative ease of maintaining a fact base rather than a rule base; the promotion of a consistent decision making strategy, and the separation of the strategy from medical details, thereby encouraging assessment and criticism inside and outside the development team. Research on the OSM is extending this approach. Methods for decision making, treatment planning and user interaction are to be dynamically constructed out of generalised components, with the specific details of different types of task stored declaratively in a set of task theories.

CURRENT STATUS OF THE PROJECT AND FUTURE WORK

In this paper we have tried to give a brief introduction to our concept and design of the Oxford System of Medicine. The prototype system which was demonstrated in September 1986 was the result of six months of development aimed at establishing whether a publication like the OSM can realistically be regarded as ripe for development, or whether it is still a goal requiring further research. The prototype contains some 3000 facts stored in an intelligible and, we believe, maintainable pseudo-English form. These lay down an overall organisation for the medical knowledge base and cover, in some detail, the information required to prescribe about forty drugs, and the knowledge required for superficial diagnosis, investigation and treatment of joint pain, nausea, breathlessness and other aspects of terminal care. The system can assist with gathering information for these decisions, offer an opinion if required, and account for its conclusions.

The aim of our study was not to produce a practical system, but to ask whether an Oxford System of Medicine is possible and, if so, to demonstrate one realisation of the concept. Although many questions remain the study has led us to the conclusion that an OSM is technically feasible and it, or systems like it, could be successfully built. The project is continuing with the establishment of an editorial board, but with the primary intention of carrying out research rather than practical development at this stage.

CAVEATS

We have tried to provide an interface that is both easy to understand and operate, leaving control of what information to provide, and when to provide it, in the user's hands. Ultimately, responsibility and authority for patient management remain with the doctor, and we have tried to avoid deliberately or inadvertently imposing any other regime. A difficulty, however, is that considerable investment is required to find out whether one is even on the right lines. Whether our design, or even a much more developed version of it, is compatible with the complexity, brevity and unpredictability of a general practice consultation ("six minutes per patient" is the UK statistic) will only be certain when it is tried in practice. Evaluation of the effect on patient care will naturally be required; although clinical audit is well established there is little experience in assessing this type of system in general practice.

Our knowledge base of 3000 facts is quite substantial but represents only a fragment of a first edition OSM - perhaps a million facts or more would be needed as a minimum. Within a few years small computer technology will probably be able to support a system on this scale, but the editorial task of expanding the system will be very great, however, and although the separation of factual material from programs promises simplified development and maintenance no one has yet built a decision support system on the scale of the OSM. The size of the effort required and the problems of scaling up are largely unknown.

The idea that generalised methods for medical decision making based on qualitative logic can be developed is a natural, though controversial, consequence of the separation of factual medical knowledge from decision making and user interaction procedures. Medicine tends to be organised and taught around experience, not generalised principles of decision making. Many may doubt the suggestion that methods for diagnosis, investigation, treatment etc can be separated from the details of what you are diagnosing, investigating or treating. If they are correct, and specific medical fields require individual problem solving techniques this will have serious implications for the viability of the OSM. The whole knowledge base would require painstaking maintenance by people with programming skills as well as medical editors. The development of an OSM would almost certainly be impractical. Although our experience is too limited to draw a firm conclusion our evidence to date is that medical knowledge can be articulated within a generalised logical framework.

NOTES

1. The "facade" seen by the user is easily altered to reflect different views of the OSM, without substantially altering the internal organisation or facilities. Other versions have had both simpler and more complicated facades. 2. Although the pseudo English form gives the appearance of being a natural language this is rather deceptive; it is still a programming language with a very specific syntax. The question of how far medical authorship can be divorced from a deeper knowledge of the working of the system remains. 3. As remarked it may in future be shown to be desirable and possible to calculate the relative preference for the decision options more precisely. Thus we might calculate the probability of each diagnosis precisely, or the expected information yield of a test or the expected utility of a treatment. Our scheme simply adds up the pros and cons, but if mathematical weights could be validly attached to the pros and cons then more precise calculations could be made without changing the overall logical framework.

ACKNOWLEDGEMENTS. Saki Hajnal for the design and implementation of the window-based interface, and for her critical and constructive comments. Tim Duncan for much of the work on Props 2.

REFERENCES

Barber D and Fox J "FIRST-AID: A design philosophy and a program for on-line symptom processing", International Journal of Biomedical Computing, 12, 249-265, 1981.

Buchanan B G and Shortliffe E H "Rule based expert systems : the MYCIN experiments of the Stanford Heuristic Programming Project", Addison Wesley, 1984.

Fitter M J and Cruickshank P J "Doctors using computers: a case study" in M E Sime and M J Coombes (eds) Designing for Human Computer Communication. London: Academic Press, 1983.

Fox, J, "Formal and knowledge-based Methods in decision technology" Proceedings of 9th International Conference on Subjective Probability, Utility and Decision-making, Groningen, The Netherlands, 1983, Acta Psychologica, 1984, 56, 303-331.

Fox, J "Knowledge, decision making and uncertainty" in W Gale (ed) Artificial Intelligence and Statistics. Reading Massachusetts: Addison Wesley, 1986.

Fox J, Barber D, Bardhan K D "Effects of on-line symptom-processing on history-taking and diagnosis - a simulation study", International Journal of Biomedical Computing, 10, 151-163, 1979.

Fox, J, Barber, D, Bardhan, K D, "Alternatives to Bayes? A quantitative comparison with rule based diagnostic inference" Methods of Information in Medicine, 19, 4, pp 210-215, 1980.

Fox, J and Frost, D "Artificial intelligence in primary care" International Conference on Artificial Intelligence in Medicine, Pavia, Italy, Amsterdam: North-Holland, 1985.

Fox J and Rector A "Expert systems for primary medical care?" Automedica, 4, 123-130, 1982.

Frost D, Fox J, Duncan T D, Preston N "Knowledge engineering through knowledge programming: the Props 2 package" Imperial Cancer Research Fund, 1986

Pritchard, P "The Oxford System of Medicine: report on a feasibility study" Oxford University Press, September 1985.

Clinical Applications (2)

EVALUATING THE PERFORMANCE OF ANEMIA

S. Quaglini(*), M. Stefanelli(*), G. Barosi(**), A. Berzuini(***)
* Dept. of Informatica e Sistemistica, University of Pavia
** I.R.C.C.S. Policlinico S. Matteo, Pavia
*** Ospedale Maggiore, Milano

INTRODUCTION

This paper reports the results of an evaluation study of the current level of performance given by ANEMIA [3], a knowledge-based consultation system addressing the clinical problem of managing anemic patients. ANEMIA has been built using an AI programming scheme, called EXPERT, which was developed at Rutgers University [6]. At present the system is able to provide assistance in the diagnosis and management of 65 disease entities. They include iron deficiency anemias, anemias due to chronic disorders, thalassemias, hemolytic anemias, arigenerative anemias and few other miscellaneous conditions.
After an extensive testing of accuracy, completeness and consistency of the knowledge-base included into ANEMIA [2,3], we designed to evaluate whether the system is able to appropriately mirror also the reasoning of well-known hematologists different from those who provided the knowledge. We were interested in testing whether there are conflicting opinions within hematologists. Implicit in the issues of performance and conflicting expertise is the question of defining a *gold standard* against which to measure the system's performance. Medical decisions may involve a great deal of subjective judgement and there may also be a great practice variation. Thus, we designed a validation study where ANEMIA's performance has been compared with that of six hematologists and the inter-expert agreement was evaluated for obtaining a realistic index of performance.

GENERAL GOALS OF THE EVALUATION STUDY

The design of the evaluation study of ANEMIA was focused on the following issues that collected data should allow to deal with.
a) The study must perform a sort of Turing Test, that is it should be able to establish if the performance of ANEMIA may be distinguished from that of an expert hematologist, when compared by a neutral observer. This kind of study will not answer the question "does the system perform the correct diagnosis ?", but the question "can the system diagnostic performance be distinguished from that provided by expert hematologists ?". In fact the first question may be meaningless, because the *correct* diagnosis is often unknown, or difficult to be proved.
b) As pointed out by Yu [7], "A complex reasoning program must be judged by the accuracy of its intermediate conclusion as well as its final decision." In agreement with this opinion, the present study was intended to evaluate also the reasoning steps of ANEMIA with respect to those followed by hematologists.
c) Since, like many areas of medicine, diagnostic task in hematology is characterized by a variety of approaches, expert hematologists may disagree among themselves. For this reason it is useful to consider the opinion of several experts on the same case.
d) Any bias deriving from a behaviour in favor or against *the computer* has to be eliminated by a "blind" procedure in which an observer judges two performances without knowing if they were produced by ANEMIA or by a hematologist.

e) Validation is often a time expensive work, especially when it involves experts from different institutions and countries. The study must therefore be designed to use their time efficiently.

THE EVALUATION PROTOCOL

The major feature investigated in the present evaluation study was the diagnostic performance of ANEMIA, distinguishing *accuracy of diagnosis* from ability to *organize the reasoning.*
The evaluation protocol we developed represents a slightly modified version of that proposed by Chandrasekaran [1]. This is described below step by step.
1- Two leading hematologists, of the Pavia University Hospital, who did not take part in the development of ANEMIA, selected 30 cases. They were only requested to form a sample of cases representative of the most frequent types of anemia.
2- Six hematologists, who were never involved in the development of ANEMIA, received, each one, the complete list of data collected in 5 out of the 30 cases. This group of experts was composed by six european hematologists, working in different institutions in different countries. The identity of each one was hidden to the others. They were requested to state a diagnosis, choosing it among those belonging to the taxonomy of diagnostic concepts included into ANEMIA (65 types of anemia), or proposing a new diagnostic concept, not known to ANEMIA. Together with their choice, they indicated a confidence level, using the following rating categories: *suspect, plausible* or *definite.* Human experts were further requested to fill a questionnaire aimed at investigating their reasoning and agreement with the organization of the knowledge base of ANEMIA.
3- The same cases were interpreted by ANEMIA, which expressed its belief in the suggested diagnosis according to the three confidence level above specified. ANEMIA also is able to justify its reasoning, as described in [2,3].
4- The diagnostic performances of hematologists and ANEMIA were arranged in a format such that the identity of the author was not recognizable.
5- Each of the six hematologists, after providing the diagnosis in 5 cases, received the two forms, one describing the hematologist's performance and the other ANEMIA's performance, for each of the remaining 25 cases. The hematologists evaluated the two performances using the following rating categories: unacceptable (i.e. *the conclusion reported in the form represents a major violation of the common medical knowledge*) , weakly acceptable (i.e. *the conclusion reported in the form is wrong, but acceptable if it was reached by a clinician non-expert in hematology),* acceptable (i.e. *the conclusion of the expert would have been different, but he considers it as one among the reasonable alternatives),* ideal (i.e. *the expert is in perfect agreement with the conclusion reported in the form).* The experts gave a separate rating to diagnoses and underlying reasoning.
6- Collected data were organized into tables that underwent a statistical analysis, as discussed later.
Before describing the results of the evaluation study, it is worth stress that although ANEMIA is able to reason on 65 types of anemia, only 30 cases were used in this study. Thus, the validation did not cover the whole domain of supposed ANEMIA's competence. On the other hand the employed methodology required that, indicated with NE the number of experts and NC the number of cases, each expert analyzed NC/NE cases, in the first phase, to provide a diagnosis and 2*(NC-NC/NE) cases, in the second phase, to judge the performances of

ANEMIA and colleagues. Taking NC=30, each hematologist reviewed 55 cases. This represents an upper limit to the effort that can be required to the hematologists involved in this study.

STATISTICAL METHODS FOR RESULTS ANALYSIS

When summarizing surveys and analysing the results of questionnaires, data consist of *counts* and *frequency tables* or *cross-tabulations* represent the common practice for their organisation.
Data collected in this study were organized into tables where row variable represents the rating given to a diagnostic performance (admissible value are *unacceptable, weakly acceptable, acceptable, ideal*) and column variable the author of the performance (admissible values are ANEMIA or hematologists). Row variable is ordinal, while column variable is nominal, but having only two levels, it can be treated as an ordinal one. Dealing with ordinal variables, tau b, tau c, gamma or Somer's D measures of association are appropriate tests. In this paper, only a brief description of the measures of association will be given. Let us consider a typical pair of observations (ratings given to a pair of diagnostic performances together with respective author), one belonging to cell (i,j) (rating i and author j), and the other to cell (i',j'). The ordinal measures of association are all simple functions of the following four quantities, computed on the count tables.

S= total number of pairs of observations for which
             $(i>i'$ and $j>j')$ or $(i<i'$ and $j<j')$
D= total number of pairs of observations for which
             $(i>i'$ and $j<j')$ or $(i<i'$ and $j>j')$
Ta= total number of pairs of observations for which $i=i'$
Tb= total number of pairs of observations for which $j=j'$
m= the minimum between the number of rows and columns
N= total number of cases

When there is strong association between the row and column variables S will be large and D will be small. Therefore, it is natural to concentrate on the size of the difference, S-D. The following statistics derive:

    a) *Goodman and Kruskal's Gamma*
       Gamma= $(S-D)/(S+D)$

    b) *Somer's D*
       D= $(S-D)/(S+D+Tb)$

    c) *Kendall's Tau*
       tau-b= $2(S-D)/[(S+D+Ta)(S+D+Tb)]^{(1/2)}$

    d) *Kendall's Tau-c*
       tau-c= $2m(S-D)/[(N*N)(m-1)]$

For a detailed description of these statistics see [5].

RESULTS OF THE VALIDATION STUDY

Individual tables were built reporting the count of ratings given by each hematologist to the performances of ANEMIA and his colleagues, while *summary* tables were obtained by summing individual tables. Denoting by El...E6 the hematologists involved in the study, Tab. la-lf show how hematologists evaluated diagnostic performances, while Tab. 2a-2f show how underlying developed reasoning was evaluated. Tables 3-4 summarize results reported in Tables 1-2.
The package SPSS/PC+ [4] was used to perform statistical analysis.

INDIVIDUAL TABLES

Tab. 1 a – Diagnoses evaluated by E1

rating counts given to

	ANEMIA	Hematologists
unacceptable	0	0
weakly acceptable	2	1
acceptable	10	4
ideal	13	20

chi – square : 4.39   signific. : 0.11
Tau – c   : 0.27   signific. : 0.02
Gamma   : 0.53
Somer's D   : 0.29

Tab. 1 b – Diagnoses evaluated by E2

rating counts given to

	ANEMIA	Hematologists
unacceptable	3	2
weakly acceptable	3	3
acceptable	5	6
ideal	14	14

chi – square : 0.29   signific. : 0.9617
Tau – c   : 0.02   signific. : 0.4401
Gamma   : 0.036
Somer's D   : 0.02

Tab. 1 c – Diagnoses evaluated by E3

rating counts given to

	ANEMIA	Hematologists
unacceptable	2	2
weakly acceptable	8	1
acceptable	8	13
ideal	7	9

chi – square : 6.88   signific.: 0.08
Tau – c   : 0.23   signific.: 0.07
Gamma   : 0.32
Somer's D   : 0.19

Tab. 1 d – Diagnoses evaluated by E4

rating counts given to

	ANEMIA	Hematologists
unacceptable	0	1
weakly acceptable	1	2
acceptable	8	5
ideal	16	17

chi – square : 2.05   signific.:0.56
Tau – c   : 0.008   signific.:0.48
Gamma   : 0.016
Somer's D   : 0.008

Tab. 1 e – Diagnoses evaluated by E5

rating counts given to

	ANEMIA	Hematologists
unacceptable	1	1
weakly acceptable	2	2
acceptable	8	7
ideal	14	15

chi – square : 0.1   signific.: 0.99
Tau – c   : 0.035   signific.:0.4
Gamma   : 0.06
Somer's D   : 0.03

Tab. 1 f – Diagnoses evaluated by E6

rating counts given to

	ANEMIA	Hematologists
unacceptable	0	2
weakly acceptable	3	0
acceptable	11	11
ideal	11	12

chi – square : 5.04   signific.:0.16
Tau – c   : 0.04   signific.:0.37
Gamma   : 0.08
Somer's D   : 0.04

INDIVIDUAL TABLES

Tab. 2 a – Reasonings evaluated by E1

rating counts given to

	ANEMIA	Hematologists
unacceptable	0	1
weakly acceptable	8	3
acceptable	7	7
ideal	10	14

chi – square : 3.9   signific. : 0.26
Tau – c   : 0.19   signific. : 0.1
Gamma   : 0.29
Somer's D   : 0.16

Tab. 2 b – Reasonings evaluated by E2

rating counts given to

	ANEMIA	Hematologists
unacceptable	5	3
weakly acceptable	4	5
acceptable	8	7
ideal	8	10

chi – square : 0.9   signific. : 0.825
Tau – c   : 0.099   signific. : 0.26
Gamma   : 0.136
Somer's D   : 0.08

Tab. 2 c – Reasonings evaluated by E3

rating counts given to

	ANEMIA	Hematologists
unacceptable	3	1
weakly acceptable	8	3
acceptable	7	13
ideal	7	8

chi – square : 5.14   signific.: 0.16
Tau – c   : 0.22   signific.: 0.07
Gamma   : 0.31
Somer's D   : 0.19

Tab. 2 d – Reasonings evaluated by E4

rating counts given to

	ANEMIA	Hematologists
unacceptable	0	2
weakly acceptable	2	3
acceptable	6	11
ideal	17	9

chi – square : 6.13   signific.:0.1
Tau – c   : − 0.34   signific.:0.01
Gamma   : − 0.53
Somer's D   : − 0.31

Tab. 2 e – Reasonings evaluated by E5

rating counts given to

	ANEMIA	Hematologists
unacceptable	0	3
weakly acceptable	6	4
acceptable	8	9
ideal	11	9

chi – square : 3.65   signific.: 0.3
Tau – c   : − 0.11   signific.:0.24
Gamma   : − 0.16
Somer's D   : − 0.09

Tab. 2 f – Reasonings evaluated by E6

rating counts given to

	ANEMIA	Hematologists
unacceptable	0	4
weakly acceptable	3	1
acceptable	11	8
ideal	11	12

chi – square : 5.5   signific.:0.13
Tau – c   : − 0.03   signific.:0.42
Gamma   : − 0.04
Somer's D   : − 0.02

Statistics reported in Tab. 1-4 do not indicate any significant difference, except for Tab.1a, where hematologists' diagnostic

Tab. 3
summary table for diagnosis

	ANEMIA	Hematologists
unacceptable	6	8
weakly acceptable	19	9
acceptable	50	46
ideal	75	87

chi – square	: 4.91	p = 0.18
Tau – c	: 0.08	p = 0.07
Gamma	: 0.15	
Somer's D	: 0.01	

Tab. 4
summary table for reasoning

	ANEMIA	Hematologists
unacceptable	8	14
weakly acceptable	31	19
acceptable	47	55
ideal	64	62

chi – square	: 5.17	p = 0.16
Tau – c	: 0.0005	p = 0.49
Gamma	: 0.0005	
Somer's D	: 0.0004	

performance was judged more favourably than that of ANEMIA, and Tab. 2d, showing that ANEMIA's reasoning was judged more favourably than that of hematologists. Direct inspection of the cell counts, disregarding statistical tests, tend to indicate a slightly higher agreement of hematologists versus their colleagues than versus ANEMIA. The opposite result was derived considering medical reasoning. This could be due to a kind of *laziness* of hematologists in justifying their diagnostic process.

Let's now give some summary statistics on judgment given to ANEMIA :
Considering the ratings *unacceptable* and *weakly acceptable* as "poor" and *acceptabe* and *ideal* as "satisfactory", 26 cases out of 30 received a number of satisfactory ratings larger than poor ones (at least 3 hematologists out of 5 judged satisfactory the ANEMIA's performance). The remaining 4 cases will be discussed later in detail.
Moreover, it is worth investigating the inter-expert consensus. In the present study, each hematologist formulated a diagnosis in 5 cases, and then reviewed the diagnostic performances of ANEMIA and 5 colleagues on the remaining 25 cases. Therefore the consensus refers to how closely different clinicians agreed in rating diagnostic performances of ANEMIA or other colleagues.
As far as judgements given to ANEMIA concern, it can be stressed that only in 3 cases a total agreement was found (5/5), in 7 cases the agreement was 4/5, in 16 cases 3/5 and in 4 cases 2/5. In 11 cases the disagreement was noticeable, since the lowest and highest rating

Tab. 5. Cell counts are the percentages of cases in which Ei and Ej were in full agreement on the basis of ratings given to ANEMIA (a) and to colleagues (b).

(a)

	E1	E2	E3	E4	E5	E6
E1						
E2	35					
E3	30	45				
E4	65	55	35			
E5	55	45	40	45		
E6	60	30	45	55	50	

(b)

	E1	E2	E3	E4	E5	E6
E1						
E2	65					
E3	30	50				
E4	80	45	35			
E5	50	40	35	60		
E6	55	40	45	45	45	

differed for two or three categories. Table 5a shows the agreement of each hematologist versus their colleagues in judging ANEMIA. Table cells report the percentages of 20 cases (this is the number of cases common to each hematologist pair), in which a full agreement was found between the two hematologists. An important disagreement was also found considering how hematologists evaluated their colleagues, as shown in Tab. 5b. These results point out a lack of large inter-expert consensus.

This supports our choice of involving several judges, since objective methods were not available to establish a diagnosis.

In conclusion, aiming at ranking competitors, that is ANEMIA versus hematologists, Tab. 6 was built: it reports the sum of the ratings given to each competitor, both for diagnosis and reasoning performance.

Tab. 6 – Ranking ANEMIA versus hematologists

	ratings sum	
	diagnosis	reasoning
E1	37	24
E2	69	64
E3	69	52
E4	62	57
E5	64	61
E6	57	54
ANEMIA	61	58

ranking on diagnostic performance : E2 = E3, E5, E4, ANEMIA, E6, E1

ranking on reasoning performance : E2, E5, E4, ANEMIA, E6, E3, E1

Tab. 6 shows that ANEMIA performs at an expert level.

UNDERSTANDING "POOR" ANEMIA'S PERFORMANCES.

An analysis of the cases in which ANEMIA's performance received the greatest disapproval can help us in discovering and correcting some deficiencies in the knowledge base or in understanding the major differences between human and ANEMIA diagnostic processes. Table 7 shows the difference between ANEMIA's and hematologist's diagnosis in the 4 cases where ANEMIA received the lowest ratings.

Tab. 7 – The four cases in which ANEMIA's performance was judged unsatisfactory

ANEMIA diagnosis	hematologist diagnosis
Anemia due to hypothyroidism (plausible) Aplastic Anemia (suspect) Congenital dyserithropoietic anemia (suspect) ratings : 0, 1, 1, 2, 3	Anemia due to hypothyroidism (plausible) ratings : 1, 2, 2, 3, 2
Acute post – hemorragic anemia (plausible) Macromegaloblastic anemia (suspect) ratings : 1, 1, 2, 2, 1	Acute post – hemorragic anemia (definite) ratings : 3, 3, 3, 3, 3
Arigenerative anemia (plausible) ratings : 1, 0, 1, 2, 1	Autoimmune hemolytic anemia due to warm antibodies (plausible) ratings : 2, 3, 3, 3, 2
Neoplastic infiltration (definite) ratings : 2, 1, 1, 1, 3	Aplastic anemia (suspect) Neoplastic infiltration (suspect) myelodisplastic syndrome (suspect) ratings : 3, 3, 1, 3, 0

In the first two cases, ANEMIA achieved the same diagnosis made by the hematologist, but suggested as *suspect* other disease entities. This kind of multiple diagnosis has been disapproved. The same occured in other 2 cases, even if higher ratings were given to ANEMIA. This could be due to the fact that hematologist used a kind of "Occam rasor": he tried to find a single diagnosis to account for the data. Since he felt to succeed, he did not indicate any other possible diagnosis. On the contrary, ANEMIA displayed all possible diagnosis suggested by data, even if at a lower level of certainty.

In the third case, a deficiency of the knowledge base was the cause of the poor performance. ANEMIA does not consider a low reticulocyte count (as it was found in this case) as a plausible finding in a case of hemolytic anemia. On the other hand, ANEMIA does not establish a diagnosis of hemolytic anemia only on the basis of a positive Coombs test, as the expert did. The deficiency of the knowledge base revealed by this case was that ANEMIA did not include into its diagnostic entities something like "Autoimmune hemolytic anemia with low reticulocyte count". This case was useful to refine the rule set for establishing "Hemolytic Anemias".

Examining the fourth case, hematologist observed some unconsistency among data and he was not able to focus on a single diagnosis. The lack of such a critiquing behavior shown by ANEMIA was disapproved.

CONCLUSION

ANEMIA's overall performance was judged acceptable in 87% (26/30) of the cases. A slightly higher overall performance, 90% (27/30), was achieved by hematologists. However, important insights could be gained analysing in detail the reasons for the small number of ANEMIA's interpretations which were judged unacceptable. The broader question of overall acceptability of ANEMIA is not addressed by this study, since we have installed only recently a version of the system running on a Personal Computer at a limited number of sites for experimental use.

AKNOWLEDGEMENT

This work has been funded by a CNR grant No.85.02630 and a MPI grant.

REFERENCES

[1]. Chandrasekaran,B. "On evaluating AI systems for medical diagnosis", The AI Magazine, 34-38, 1983.

[2]. Cristiani,P., Quaglini,S., Stefanelli,M., Barosi,G., Berzuini, A. "MICROANEMIA: An Expert System Running on a Microcomputer", Proceedings MEDINFO '86, 87-91, Washington 1986.

[3]. Quaglini, S. and Stefanelli,M, Barosi,G. and Berzuini,A. "ANEMIA: An Expert Consultation System", Computers And Biomedical Research, 19, 13-27, 1986.

[4]. SPSS Inc., M.J.Norusis, SPSS/PC + for the IBM PC/XT/AT.

[5]. Upton G.J.G. The Analysis of Cross-tabulated Data. John Wiley & Sons Ed., 1978.

[6]. Weiss, S, and Kulikowsky,C. "EXPERT: a system for developing consultation models", Proc. Int. J. Conf. Artif. Intell. 4th, pp 841-846, Tokyo, 1979.

[7]. Yu,V.L., Buchanan,B.G., Shortliffe, E.H., Wraith, S.M., Davis,R., Scott,A.C. and Cohen,S.N. "Evaluating the Performance of a Computer-based Consultant", Computer Programs in Biomedicine, 9, 95-102, 1979.

<u>**COMPUTER AIDED DIAGNOSIS AND TREATMENT OF BRACHIAL PLEXUS INJURIES**</u>

R.B.M. Jaspers,[1,2] F.C.T. van der Helm[3]
Delft University of Technology, Fac. of Mechanical Engineering,
Lab. for Measurement and Control, Man Machine Systems Group.

ABSTRACT

The expert system PLEXUS establishes a differential diagnosis for brachial
plexus injuries. Ultimately, it will also present a treatment plan and
prognosis for a specific patient, based upon his/her diagnosis. The
diagnostic part of the system has been implemented in the rule based expert
system shell Delfi-2. In order to deal with the practically unlimited number
of possible diagnoses, which are combinations of various injuries, a
classification of evidence has been introduced that facilitates the
evaluation of each injury. In this way a transparent system is obtained. For
the major subset of injuries, i.e. the supraclavicular injuries, the results
prove to be quite good. Further research is conducted to improve the
diagnostics of infraclavicular injuries and to implement the prognostic and
treatment planning module.

INTRODUCTION

Injuries to the brachial plexus particularly occur to young men from motor
cycle accidents. But many other causes of injury exist, like stabwounds or
shotwounds, compression and post-radiation fibrosis. In general the
consequent loss of function in the arm and hand, will have considerable
implications for the life and future plans of these patients. Therefore the
treatment of brachial plexus injuries not only concentrates on somatic
functional disturbances, but it comprises a complex of activities in the
medical, paramedical, occupational, technical, social and psychological
fields. Because of this a multidisciplinary approach is necessary for an
optimal treatment (Van Haselen 1985). Often associated injuries occur, like
head trauma, fractures to the trunk and extremities or vascular lesions.
These may complicate the treatment very much. Therefore the average brachial
plexus treatment process takes about 3 years (Jaspers 1982). The complexity
of the treatment process makes it hard for the treatment team to plan an
optimal treatment for a specific patient.

TREATMENT OF BRACHIAL PLEXUS INJURIES

The treatment of brachial plexus injuries largely depends on the severity
and the location of the injury. Mild injuries like neurapraxia and
axonotmesis will generally show good recovery, so these do not need to be
treated surgically. Severe injuries (neurotmesis) often require

---

The research reported here has been partially supported by:
1. The Delft University Fund
2. The Prevention Fund
3. The Dutch Ministry of Social Affairs.

neurosurgical repair for a fair prognosis, although injuries causing root avulsions only have limited possibilities for surgical repair. Globally, dependent on the patient's diagnosis, the treatment scheme should consist of the following:

In the early stage: − Diagnostics.
           − Conservative treatment.
           − Neurosurgical repair, if necessary.

In the middle stage: − Multidisciplinary conservative treatment.

In the late stage: − Reconstructive orthopedic surgery, if necessary,
           Followed by conservative treatment.

In this treatment scheme it is of utmost importance to diagnose the brachial plexus injury as early as possible, since the assessment for neurosurgical repair has to be done within 4 months after the injury. After this period the prognosis of this treatment gets much worse (Narakas 1985, Millesi 1980, Sunderland 1978). Early diagnosis and neurosurgical repair may also lead to perform the reconstructive surgery at an earlier time, which will benefit the rehabilitation.
As will be clear from the foregoing, a sufficient diagnosis of the brachial plexus injury is vital for the planning of the treatment for a specific patient. However, in a great number of patients the diagnosis is not sufficiently known to establish their treatment plan. For instance, from the patients that are referred to one of the major Dutch rehabilitation centers, 45% percent appears to have a not sufficiently diagnosed brachial plexus injury (Jaspers 1986). This is due to the fact that:

- In the early stage often associated injuries ask for attention, so that the brachial plexus injury is left unattended.
- The diagnostics of brachial plexus injuries is a very complex process, since very many different kinds of injury exist, and in one patient several injuries may exist at the same time.
- In a great number of cases (20%) in the hospital no diagnostic tests are performed, apart from clinical evaluation.
- Often the results of diagnostic tests are not available, are unclear or incomplete.
- Often the results of several diagnostic tests, performed at different places, are not combined to a sufficient diagnosis of the injury.
- In many hospitals there seems to be a lack of knowledge about the possibilities of modern treatment schemes, e.g. neurosurgical repair, for the treatment of these injuries.

These problems are caused by a lack of experience with the diagnostics and treatment of brachial plexus injuries due to the comparatively small number of these injuries that is treated in most Dutch hospitals. In only two hospitals specialized treatment teams operate that have a sufficient experience with the diagnostics and neurosurgical repair of brachial plexus injuries. Therefore it was decided to develop a diagnostic and treatment support system for brachial plexus injuries. This system may be used to propagate brachial plexus diagnostics and treatment expertise to other hospitals.

THE TREATMENT TEAM SUPPORT SYSTEM 'PLEXUS'

The treatment team support system PLEXUS is aimed at the improvement of the diagnostics of brachial plexus injuries and at the propagation of the knowledge about the possibilities of modern treatment schemes. In order to achieve this the system should not only aid the diagnostics, but also it should be able to present a treatment plan and prognosis based upon this diagnosis. It should be appreciated that this is not the way experts work. In general, although he is capable of doing this pre-operatively, a brachial plexus expert will carry out the diagnostics as far as necessary for the assessment of the patient for neurosurgical repair. Then per-operatively a more detailed diagnosis and treatment plan will be determined. It was felt, however, that for the Plexus system to be acceptable by it's potential users, it should not only do the assessment for neurosurgical repair, but it should also be able to give a prognosis for this treatment and it's alternatives. In this way it is taken care of that no new treatment strategy is imposed on the physician without being able to explain why this might be the optimal treatment for a particular patient. For this purpose the PLEXUS system consists of three different modules:

1. A brachial plexus diagnostic module.
2. A brachial plexus treatment planning module.
3. A brachial plexus prognostic module.

Because of their widespread applications that seem to prove the suitability of this concept in the medical field, expert systems are used for the implementation of these modules. The expert system PLEXUS has been based upon the rule based expert system shell Delfi-2 (De Swaan Arons 1984, Lucas 1987). This shell has a backward chaining inference engine and is equipped with the certainty factor method for probabilistic reasoning (Shortliffe 1976). At this moment the implementation of the diagnostic part of the system has been practically finished, whereas the other two parts are still being worked on. Therefore, only the diagnostic module will be discussed in some more detail.

The knowledge that has been implemented in the diagnostic system has been acquired from two experts in the domain of the diagnostics and neurosurgical repair of brachial plexus injuries (Thomeer 1986, Slooff 1985, 1986). This domain has a number of important properties:

- In general the diagnosis of a brachial plexus injury will be a combination of injuries that are present at the same time. The number of possible injuries is about 50, but the number of combinations of these is practically unlimited.
- Facts that can be established by diagnostic tests will generally point to several injuries. Therefore it is hard to make a differential diagnosis of the brachial plexus injury.
- A large number of diagnostic tests is available, which will not always give reliable information. The validity of these tests may depend on the results of other tests. Moreover, the evidence acquired from these tests is often hard to combine to a differential diagnosis, because the results are not always in accordance with general theory. For instance, individual variations in the anatomy of the brachial plexus may hamper the diagnostics.

These domain properties present a number of problems in establishing a differential diagnosis for a specific patient:

- Necessary tests may not be done for a specific patient.
- Tests may be done that are not valid for a specific patient.
- Test results are easily misinterpreted because of the frequent individual variations in the anatomy of the brachial plexus.
- To combine test results to a differential diagnosis for a specific patient is hard because of the large number of possible diagnoses and because a particular fact may point to various injuries.

To assure that a good diagnosis is obtained by the PLEXUS system, it had to be taken care of that the right diagnostic tests are performed. This presents the necessity for the system to not only present it's conclusions about the site and severity of the injury, but also to advise the user as to what supplementary diagnostic tests are necessary to improve the reliability of the diagnosis. Furthermore, to prevent the performance of invalid tests for a particular patient, the PLEXUS system has to be able to point out imperfections in the diagnostic procedure taken by unexperienced users. These features have been implemented in the system.
Concerning the management of uncertain information, two points are of interest. First of all the Delfi-2 shell provides the certainty factor method for this. An important source of uncertainty, i.e. the occurrence of inconsistent evidence, can not be dealt with in this way. In the PLEXUS system this has been solved as follows: When the occurrence of inconsistent evidence is traced by the system, it will reason about the justification for this evidence by means of additional information. In this way, often conclusions about the validity of the evidence can be achieved. If this is not possible, however, the PLEXUS system will come to a conclusion by default. Then, it will also inform the user about this and call attention to the discrepancies in the test results.
Another important point about the processing of uncertain information is that large amounts of evidence may fit in with various diagnoses. This problem has been solved in a way that highly resembles the expert's intuitive line of reasoning. For each diagnosis, all evidence has been classified into four categories:

1. Necessary facts. For a specific diagnosis it's necessary facts have to be present in order to be able to conclude that diagnosis.
2. Contra-indications. In order to be able to conclude a certain diagnosis, it's contra-indications may not be present.
3. Corresponding facts. Once a diagnosis has been established, certain facts that conform to this diagnosis are used to raise it's certainty.
4. Irrelevant facts. For a specific diagnosis, certain facts may be irrelevant.

Because of this classification, a certain piece of evidence may be a necessary fact for a number of diagnoses as well as a contra-indication for others. An example of this is shown in Table 1.
In order to establish a certain diagnosis, a number of so-called necessary facts must be present and no contra-indications may exist for that particular diagnosis. If for a certain diagnosis these conditions have been fulfilled, the certainty factor of this diagnosis can be raised by so-called corresponding facts. The following rules will elucidate this procedure for the diagnosis "avulsion of spinal nerve root C6":

Table 1. Example of the classification of evidence.

EVIDENCE / DIAGNOSIS ⟶	AVULSION C5	TRUNC. SUP.	FASC. DORSALIS
LOSS OF MOTOR-FUNCTION C5	necessary	necessary	irrelevant
NERVE ROOT C5 NOT-VISIBLE	necessary	contra-ind.	contra-ind.
TINEL RADIATES TO C5	contra-ind.	correspond.	irrelevant

IF
1. The fila radicularia of spinal nerve root C6 are not visible in the
   myelogram ;
 and
2. No contra-indication for the diagnosis 'avulsion of spinal nerve root C6'
   is present ;
THEN
1. The diagnosis of the patient is 'avulsion of spinal nerve root C6' ;
 CF=0.90

IF
1. The patient shows a Tinel's sign ;
 and
2. The location of the Tinel's sign is supra-clavicular ;
 and
3. The Tinel's sign radiates to the dermatome C6 ;
THEN
1. A contra-indication for the diagnosis 'avulsion of spinal nerve root C6'
   is present ;
 CF=1.00

Also other contra-indications can exist. If condition 1 of the first rule
has been fulfilled and if the PLEXUS system can not conclude one of the
contra-indications, then it concludes that part of the diagnosis for this
patient is 'avulsion of spinal nerve root C6'. Other rules that conclude
about the multi-valued parameter 'diagnosis' may be true also. In this way
the final diagnosis for a particular patient mostly consists of a number of
non-mutually exclusive injuries that is present at the same time (Figure 1).
By means of corresponding facts the certainty of the concluded diagnosis may
be raised:

IF
1. The diagnosis of the patient is 'avulsion of spinal nerve root C6' ;
 and
2. A meningocele of spinal nerve root C6 is visible in the myelogram ;
THEN
1. The diagnosis of the patient is 'avulsion of spinal nerve root C6' ;
 CF=0.50

```
** DELFI-2 ACHIEVEMENT SUMMARY **

Concluded  ( through rule(s)  142 151 242 150 140 ) :
the diagnosis for supraclavicular lesions of the patient is:
        1. extraforaminar-lesion-c6, with certainty:  0.94
        2. extraforaminar-lesion-c7, with certainty:  0.93
        3. partial-lesion-c5, with certainty:  0.50

Concluded  ( through rule(s)  1311 ) :
the diagnosis for infraclavicular lesions of the patient is:
        1. none, with certainty:  1.00

Concluded  ( through rule(s)  1009 ) :
the recommended treatment of the patient is:
        1. to-consider-surgical-repair, with certainty:  1.00

Concluded  ( through rule(s)  779 ) :
the examinations which are required for an exact diagnosis of the patient is
        1. check-sign-of-Tinel, with certainty:  1.00
```

Figure 1: Diagnosis achieved by the PLEXUS system for a patient with a
brachial plexus injury

As a result of this method, a transparent system is obtained, that very much
resembles the medical line of reasoning and that uses all information
relevant to evaluate a certain diagnosis. Presently the diagnostic part of
the system contains over 900 production rules.

PRELIMINARY RESULTS

A preliminary test of the system's diagnostic performance has been carried
out. In this test 32 patients with traumatic brachial plexus injuries were
diagnosed by the PLEXUS system. Of those, 26 suffered from a supraclavicular
brachial plexus injury, 17 of which had been neurosurgically reconstructed.
There were 3 patients with infraclavicular brachial plexus injuries, 1 of
which had been operated upon. And 3 patients suffered from a combined supra-
and infraclavicular injury, which had all been operated upon. All patients
had been treated in the Leiden Academic Hospital. The goal of this
preliminary test was to determine the performance of the PLEXUS system
relative to the expert's performance. Since the goal of the system is to
present a diagnosis as well as a prognosis and treatment plan, the PLEXUS'
diagnosis should be a good approximation of the expert's per-operative
diagnosis. Therefore in the preliminary test the system's diagnosis was
compared to the expert's per-operative diagnosis, whenever this was
available. In the other cases the expert's clinical diagnosis was used as a
reference. The following criteria have been used to measure the system's
performance:

- A system's diagnosis is judged 'good' if it corresponds to the expert's
  (per-operative) diagnosis.
- A system's diagnosis is judged 'fair' if it only slightly deviates from
  the expert's diagnosis. Particularly these deviations may not have an
  effect on the treatment planning and prognosis of these injuries.
- In all other cases the prognosis is judged 'poor'.

In Table 2a the results of these tests are presented. It can be seen that for supraclavicular injuries (injuries above the clavicle) the results of the system are already quite good. Generally these are the more severe injuries and they constitute the largest group of brachial plexus injuries (about 70%). Some small imperfections have yet been solved, so now it seems that these injuries are dealt with very good by the PLEXUS system (Table 2b), although a final test on a new series of patients still has to prove the actual quality. As far as the infraclavicular (below the clavicle) and combined injuries are concerned, still work is being done to improve the diagnostics of these injuries. This will be discussed in the next paragraph.

Table 2a: Preliminary results of the PLEXUS system's diagnostics.

INJURY	RESULT → OPERATED	GOOD	FAIR	POOR
SUPRA- CLAVICULAR	Yes (17) No (9)	14 (82%) 8 (89%)	2 (12%) 1 (11%)	1 (6%) 0
INFRA- CLAVICULAR	Yes (1) No (2)	1 (100%) 0	0 0	0 2 (100%)
COMBINED	Yes (3)	0	2 (67%)	1 (33%)
TOTAL	Yes (21) No (11)	15 (71%) 8 (73%)	4 (19%) 1 (9%)	2 (10%) 2 (18%)

Table 2b: Preliminary results of the improved system, based upon the same series of patients.

INJURY	RESULT → OPERATED	GOOD	FAIR	POOR
SUPRA- CLAVICULAR	Yes (17) No (9)	16 (94%) 9 (100%)	1 (6%) 0	0 0
INFRA- CLAVICULAR	Yes (1) No (2)	1 (100%) 0	0 0	0 2 (100%)
COMBINED	Yes (3)	1 (33%)	2 (67%)	0
TOTAL	Yes (21) No (11)	18 (86%) 9 (82%)	3 (14%) 0	0 2 (18%)

DISCUSSION AND FURTHER RESEARCH

As will be clear from the foregoing, the system's diagnostics of infraclavicular brachial plexus injuries is not yet acceptable for use as a treatment team support. These injuries often are comparatively mild and show a good natural recovery. Therefore, at the time the diagnostics is performed, a number of muscles may have been recovered already, which may obscure the actual diagnosis. Because of this the initial pattern of motor and sensory disturbances should be reconstructed first, before determining the site of the injury. These aspects have not yet been implemented in the PLEXUS system. The problem of processing information that is time-varying is important for other parts of the PLEXUS system also, like the prognostic part. The prognosis of brachial plexus injuries largely depends on the diagnosis as well as the initial motor and sensory function and the improvement therein over time. Another problem in the implementation of the prognostic part of the system is posed by the fact that the expert's knowledge about the prognostics is rather limited. Because of this at the Leiden Academic Hospital a detailed study of the prognosis of these injuries is presently undertaken. However, prognostics inherently has a fuzzy character. Probabilistic information about the prognosis of these injuries will not be readily available since so many different diagnoses exist and because of the relatively small number of patients. Therefore knowledge about prognostics will be in terms like: "If the diagnosis for a patient is such, and the initial muscle function is poor, and the reconstructive surgery has been done in an early stage, then this muscle may show fair recovery after about one and a half years.". Because of the intrinsic fuzziness of the prognostics, the use of fuzzy logic will be investigated for the implementation of this part of the system.
Another important aspect that is given much attention is the actual introduction of the PLEXUS system in the treatment centers. This makes large demands on the user interface and the computational efficiency of the system. It is found that the method of knowledge representation, i.e. production rules and backward chaining, presents some disadvantages in this respect. Especially, this way of knowledge representation not always corresponds to the way experts work. The determination of the site of the injury, for instance, they do in a way of pattern recognition, which is hard to implement in production rules. Therefore it has been implemented in production rules in a different way, which is less computationally efficient. Because of this, other methods of knowledge representation are under investigation, particularly the introduction of frames. For this particular problem other search strategies are tried, like best-first strategies instead of the depth-first strategy that is presently used.
Furthermore, for introduction in the treatment centers, more extensive database facilities will be built in.

Finally some disadvantages of expert systems have been ascertained that we try to overcome by means of other modelling techniques. Since the aim of expert sytems is to lay down experts' knowledge about a certain problem area, the best to be achieved is to approximate their results as well as possible. First of all this poses the problem of finding real experts in the problem domain. But also it means that, in contrast to other modelling techniques, an expert system will never present new information about the domain that was not already known by experts. This is particularly of interest for the prognostic part of the system, since knowledge is still rather limited in this field. Because of this disadvantage, ultimately in the PLEXUS system a different modelling approach will be implemented that may give new insights in the process, i.e. system identification and parameter estimation.

It is well known in medical engineering that, based on measurements on a patient, a model can be derived that calculates the most probable future state of the patient (Stassen 1980, Blom 1975). The main difficulty in this approach is the choice of input and output signals and the model structure. The quality of the model's prediction depends on this. From this model, for a specific patient a prediction of his future state (prognosis) can be made. Also from a desired outcome of the treatment process an optimal treatment plan can be calculated and new insights in the process may be gained, e.g. the effect of a certain treatment on the patient's state can be tested, without having to test this on the patient himself.

However, this approach presents some important disadvantages also. First of all the method by which conclusions are reached by such a patient model will not be clear to the user, since the model has no physiological significance. Probably this will hamper the acceptability of the model. Secondly a lot of data is required in order to derive a model, and also some important data can hardly be measured, like psychological and social variables that play an important role in the treatment process. In the PLEXUS system the usefulness of this approach is investigated in comparison to and in combination with the use of expert systems.

ACKNOWLEDGMENT

The expert help of Dr.R.T.W.M. Thomeer of the Leiden Academic Hospital and Dr.A.C.J. Slooff of the De Wever Hospital in Heerlen that made this work possible, as well as the kind help of W.B.M. Slooff are gratefully acknowledged.

REFERENCES

Blom, J.A. (1975). Trend prediction and automated therapy in patient intensive care; in: Computers in cardiology, pp.213-214, Rotterdam.

Haselen, W.E.C. van (1985). Rehabilitation of severe nerve injuries of the upper extremities; in: Symposium "Diagnosis and treatment of nerve lesions", Erasmus University Rotterdam.

Jaspers, R.B.M. et al (1982). Evaluation of the treatment process of patients with brachial plexus injury (in Dutch); in: Proc. Boerhaave Committee "Traumatic brachial plexus injuries", pp.65-89, Leiden University.

Jaspers, R.B.M. (1986). Diagnostics of brachial plexus injuries (in Dutch); Report N-259, lab. WMR&CE, Delft University of Technology.

Lucas, P.J.F.; H. de Swaan Arons (1987). Extensions to the expert system shell Delfi-2; in: 2nd Mini conference on expert systems and operations research, Reidel Publishing Company, Dordrecht.

Millesi, H. (1980). Trauma involving the brachial plexus; in: Management of peripheral nerve problems; eds. G.E. Omer, M. Spinner, pp. 548-568, W.B. Saunders, Philadelphia.

Narakas, A.O. (1985). The treatment of brachial plexus injuries; in: International Orthopaedics, vol.9, no.1, pp. 29-36.

Shortliffe, E.H. (1976). Computer-based medical consultations: MYCIN; Elsevier, New York.

Slooff, A.C.J. (1985). Some comments on diagnosis and treatment of traumatic brachial plexus lesions; in: Symposium "Diagnosis and treatment of nerve lesions", Erasmus University Rotterdam.

Slooff, A.C.J. (1986). Private communication.

Sunderland, S. (1978). Nerves and nerve injuries; 2nd edition, E&S Livingstone, London.

Stassen, H.G. et al (1980). A computer model as an aid in the treatment of patients with injuries of the spinal cord; in: Proc. of the Int. Conf. on Cybernetics and Society, pp. 385-390, Massachusetts.

Swaan Arons, H. de; P.J.F. Lucas (1984). Expert systems in an application environment (in Dutch); in: Informatie, vol.26, no.8, pp.631-637.

Thomeer, R.T.W.M. (1986). Private communication.

J.M.Goutal (1), N.Philip (2), M.Griffiths (1), S.Aymé (2)

(1) GRTC - CNRS,
31 chemin Joseph Aiguier,
13402 Marseille Cedex 9, France.

(2) INSERM U242,
Centre de génétique médicale, Hôpital de la Timone,
13385 Marseille Cedex 5, France.

1. INTRODUCTION

Medical experience shows us that major embryonic anomalies are present in 3% of births. In 15% of these cases, several anomalies are to be found. It is important to understand the mechanisms behind such associations in order to evaluate the risk of recurrence for a subsequent child produced by the same couple.

Etiologies of multiple malformations can be classified in three groups: chromosomal abnormalities, genetic syndromes and environmental defects. In the first two groups, the initial cause is present in the first cell. In the last group, the cause occurs during later development. Consequences can be common to the three groups.

It is always assumed that multiple malformations are due to a single cause. The single event can take place at a specific time, leading to anomalies in all the processes going on at that moment in time. It can be tissue specific, affecting those organs derived from the particular tissue. It can also modify a specific process which is involved in the organisation of several organs.

By describing normal embryonic development stage by stage, process by process, together with its anomalies, it should be possible to test, starting from the description of an abnormal pattern, the three hypotheses: timing, single abnormal process, or unique tissue? The same model can also be used to predict the final anomalies resulting from a given perturbation at a specific moment or in a specific tissue or process.

Exhaustive description of embryonic development would have implied thousands of rules, but in fact not all the stages could reasonably be involved in viable malformations (the only ones which interest us at this time). We have therefore, for the moment, limited our description to that of processes which can be at the origin of the most common malformations.

Some very real unsolved questions about the origin of multiple malformations could probably be answered using such a system. For instance, the so-called "VACTERL" association, which includes vertebral defects, tracheo oesophogeal fistula, anal atresia and radial defects, has no known origin, but is observed all over the world more often than could be expected from the individual incidence of each malformation. Many of these so called "associations" are puzzling for the medical geneticist who has to advise parents.

2. WHY DO WE NEED AN EXPERT SYSTEM?

Given the amount of work involved in making an expert system, a necessary preliminary is always to justify the use of artificial intelligence techniques. In this particular case, the reasons are standard. Firstly, as with other diagnostic systems, the methods used are those of logical reasoning. The use of an expert system in this field is not exactly original, see for example (Fieschi 1984). Secondly, the number of rules is extremely large.

Since the information must be validated by medical experts, it must be coded in a form which is accessible to them without imposing the learning of a programming language. Thirdly, a complete formal description of embryonic development has never been produced. The knowledge base will thus need to be debugged by the embryologists. Even if it were possible to describe this extremely complex process without error, the knowledge base would still be evolutive. This results from the fact that it will be used as a vehicle for the exploration of hypotheses by research workers, who will automatically wish to improve the description as a result of their work.

3. METHODOLOGY.

The essential characteristic of an expert system is contained in the word "expert". It models some competent person's expertise. As far as possible, computing constraints should be hidden from the expert, in order to allow him or her the maximum of freedom. Of course, the result must be a formal document, but the choice of formalism depends on the application field. The computer scientist must adapt to the user, not the other way round. Paradoxically, this can create a psychological blockage for users with computing experience. They have learnt that programs impose heavy constraints, and continually wonder whether their liberty of expression will be "understood" by the expert system.

We consider that adapting to the user means that implementation tools are chosen in function of the knowledge base, that is at the end of the process of gathering expertise. In practice, things are not as simple as this. In the project we describe, the amount of information in the knowledge base will ultimately be very large, of the order of several thousand rules. We consider that it would be a risk to create such a base in one fell swoop. A prototype will exist first, limited in size to be easily implemented in PROLOG on available equipment. From the expert's point of view, this limitation is simply one of the number of rules, the form of which is not that of the implementation language. The complete system will not be implementable in PROLOG, for reasons of size, unless, by the time we are at that stage, new interpreters become available. The problem with working in a standard programming language such as PASCAL would be that the program becomes difficult to modify. Making the system evolutive will require an expert interface.

4. INFORMATION STRUCTURES.

Two types of information are provided by the expert. Firstly, information about tissues leads to data base facts. Each tissue is described by the set of one or more tissues from which it is developped, the new tissues which are developped from it, the time step at which it is created and the biological process used. The expert can also indicate conditions, such as proximity or gender, necessary for the carrying out of a process. Note that abnormal tissues are described in the same way, but in their description further information is given concerning the cause of the abnormality. For example, this may be a non terminated process, or an insufficient supply of tissue.

Tissue development is described by a sort of grammar, in which rules have the following form:

OT1 + OT2 + ... -> NT1 (ti) NT2 (tj) (process) I (cause) NT3 ... I ...

From a set of existing tissues OT1, OT2, ... is created a set of new tissues or organs NT1, NT2, ..., at times ti, tj, ... by named processes. The first possibility is always the normal derivation. Abnormal situations follow, preceeded each time by the cause, otherwise with the same syntax. For example:

    muller_channel -> (female) uterine_ducts uterus vagina
                       (phase 23) (growth) (closing)
                       I (male) (phase 23) (regression)

This rule says that women grow their sexual organs at phase 23, by growing and closing, whereas men produce no new tissues or organs at phase 23, the muller channel disappearing by regression.

The time scale is a sequence of integers from 1 to N which indicate the stage of embryonic development. Each stage is fixed in real time, stage 23 corresponding, for example, to day 57 of pregnancy. Different stages, however, require different lapses of time.

The second type of information contains additional facts about perturbations. For instance, an illness of the mother or the description of a genetic factor and its consequences. These are described by rules which validate or invalidate facts in the tissue data base. Rules are expressed as
IF ... THEN ...   statements, as in many systems of this type.
For example:

IF hyperthermia
AND phase=X
THEN slow_growth

These formalisms are accessible to the experts, who have the difficult task of assembling all the information. In particular, they can modify and add information without consulting the computer specialists. This is one of the more important reasons for using artificial intelligence methods.

5. PROCESSING.

Once the formalisms used to describe the data base and the rules are fixed, the experts need an interface which allows them, using standard menu techniques, to add information or modify existing information. The knowledge base is a primary tool for the expert's reflection over a period of time.

Interrogation is by a similar interface, which allows questions concerning, for example, possible common ancestors of different tissues or organs, lists of tissues produced during a given time phase, as a result of a given process, ... Questions like these allow the expert to formulate hypotheses or to look for common phenomena. He can then test a hypothesis by introducing a perturbation, and running the system which simulates embryon development with it. This allows comparison of observed and simulated cases.

The development tree for tissues and organs corresponds to a standard AND/OR tree in computing. The program covers the tree in backward or forward chaining. Forward chaining simulates embryo development under imposed conditions from the egg (the root of the tree) to the final set of organs (its leaves). Backward chaining leads from a clinically observed set of malformations (or a set of test data) to the root, indicating possible causes.

6. IMPLEMENTATION.

The protype is in the process of implementation in PROLOG on an Apple MAC+. The language is useful at this level, allowing rapid testing of ideas. Heavy use is made of the backtracking mechanism, for obvious reasons. An evaluation of the protype will be necessary before we commit ourselves finally.

The Macintosh allows easy communication with the mouse, the graphic screen and its menu system. The multiview capability allows one window to be used for text (rules, for example), one for the menu and one for the problem under consideration.

7. CONCLUSION.

At the time of writing, the system is at the start of its implementation phase. Results will be available for presentation at the conference. The first implementation concerns the limited prototype. This will allow us to test the primitives chosen for the representation of the knowledge base, and to carry out experiments on the correction of the system by treating cases presently held in available medical records.

This first system will not be distributed. To make it into a complete, available system will require much work from both the medical and the computing sides of the project. Completing the knowledge base will occupy medical researchers for some time. They will need to access the system through a convivial interface without consulting computing staff. This interface is non trivial to construct, and will form the subject matter of a subsequent paper.

If the complete system is delivered as we foresee it at this moment in time, it will be one of the largest expert systems to be put into service. This factor is sufficient to make us cautious about going too fast. Testing the prototype is a necessary stage of development.

8. REFERENCES.

M.Fieschi
Intelligence artificielle en médecine - des systèmes experts
Masson 1984

R.G.Harrison
Clinical embryology
Academic Press 1978

D.W.Smith Recognizable patterns of human malformation
W.B.Saunders 1982

<u>MICRO COMPUTER BASED DECISION SUPPORT FOR LIPID DISORDERS</u>

Deirdre Ni Fhaircheallaigh (1)

Margaret Sinnott (2)

Jane Grimson (1)

Tony McGill (2)

Rory O'Moore (2)

Department of Computer Science (1) and Clinical Biochemistry (2),
Trinity College, Dublin University.

<u>INTRODUCTION</u>

The decision support system described relates to the problem of the
diagnosis and management of patients with disorders of lipid
metabolism.   Dietary and/or pharmacological intervention   in
patients with hyperlipidaemias is now widely accepted as being
beneficial. (1)   The recent Lipid Research Clinics program has
shown that energetic reduction of elevated cholesterol levels
reduces the increased risk of coronary heart disease. (2)

The interpretation of plasma lipid or lipoprotein levels is a
complex process for the non specialist physician.   For example in
addition to the inherited (primary) lipid disorders, there are
numerous drugs and other diseases which influence lipid metabolism.
Many drugs also interfere with the analytical methods used in the
measurement of serum lipids.

We believe that the lipid decision support system which takes all
of the above variables into account will prove a useful asset in
the interpretation of lipid profiles.   The system has been
developed to:-

1) provide a laboratory tool for the evaluation of the relative
   clinical and cost effectiveness of the   available biochemical
   tests.   It is hoped that this may lead to a reduction in the
   number of expensive assays required to make the appropriate
   diagnosis.

2) provide interpretative reports for users of the laboratory
   particularly general practitioners and other non specialists in
   the lipid area who are becoming increasingly involved in the
   diagnosis and management of patients with hyperlipidaemias.

3) Assist in the management of patients attending hospital lipid
   clinics.

The system also has potential as an educational aid since the
reports generated display the logic used to reach the diagnosis.

Initial assessment of patient results is by the most widely accepted general classification of hyperlipidaemias - The World Health Organisation modification of the original Fredrickson classification (Table 1). (3)

Type	Prevalence	Liproprotein abnormality	Usual changes in plasma lipid concentrations		Stored plasma test	
			chol.	trig.	top layer	infranatent
1	Rare	Chylomicrons	+	+++	cream	clear
IIa	Common	Low density(beta) lipoprotein	++	normal	nil	clear
IIb	Common	Low density(beta), very low density pre(B) lipoproteins	++	++	nil	clear or slightly turbid
III	Uncommon	'Broad-beta' (floating beta) lipoproteins	++	++	slight cream	turbid
IV	Common	Very low density (pre-beta) lipoprotein	normal or +	++	nil	turbid
V	Uncommon	Very low density (pre-beta) lipoproteins ,chylocicrons	+	++	cream	turbid

Classification of characteristics of Hyperlipidaemias
Table 1

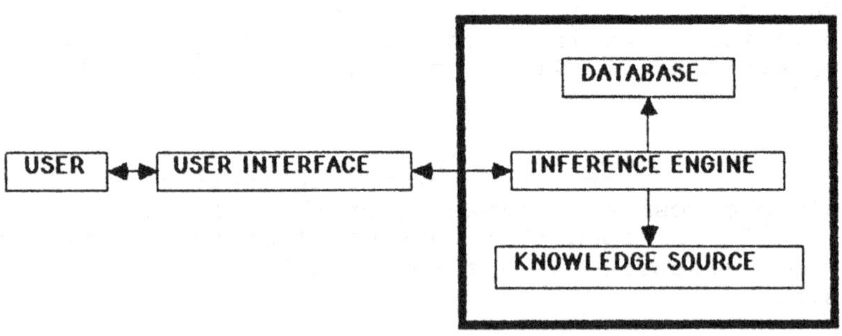

EXPERT SYSTEM DESIGN
Figure 1

THE SYSTEM

The program was originally developed in a version of LISP - Franz LISP on the university VAX 11/780 mainframe. The program has been rewritten in Golden Common LISP, a dialect of common LISP and now running on an ERGO PC 88 (IBM X T Compatible) using MSDOS (Version 2.11). The system consists of a knowledge base and an inference engine (Figure 1).

The knowledge source is represented in the form of production rules. The rules follow the format below:

RULE NAME

   (If (premise I)......(premise-N))

   (then (Comments LIST) (Action I).....(Action N))

e.g. Rule R. 15

```
        (IF        (PATIENT NOT FASTING)
                   (TOTAL CHOLESTEROL HIGH)
                   (TRIGLYCERIDE HIGH)
                                              )

        (THEN      (COMMENT 2)
                   (SKIP INTERPRETATION)
                   (CHECK DRUGS AND DISEASES)
                                              )
```

The knowledge base consists of the set of rules, tables and lists required to operate the system. It also contains the comments which form the basis of the reports. The evaluation of lipid disorders is complex due to the many variables which may influence plasma lipid levels. Thus tables of reference ranges by age and sex of the different lipid substances measured are included. Together with lists of known secondary causes of hyperlipidaemia, both drugs and diseases. These are required to modify the initial interpretation of the test results. Lists of symptoms and signs associated with the lipid disorders are also maintained and used to increase the accuracy of the diagnosis.

The inference engine (Figure 2) drives the system proceeding through the rule lists analysing the premises of each rule in turn. If the premise of the rule evaluates to true the rule is activated which then:

1.   stores comments for reports

2.   performs sub routines requested by the rule

3.   builds a list of facts relevant to the individual case

The inference engine controls the complete running of the program.

SYSTEM FUNCTIONS

The main menu offers four choices.

1) Help, 2) Supervisor mode, 3) User mode, 4) Exit.

The help module describes the system and its components. The supervisor mode is password controlled. It allows the critical decision and reference ranges (4) to be changed to suit local requirements. The syntax of the report comments may also be

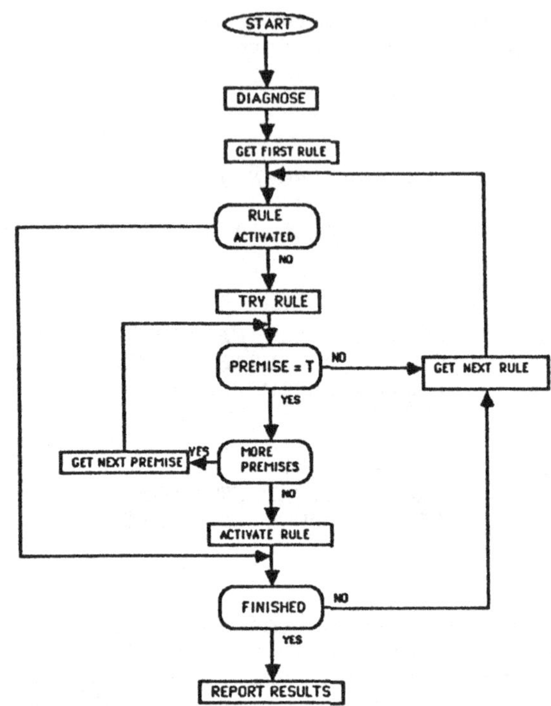

Inference Engine Flow-Chart

Figure 2.

modified. Existing biochemical tests may be replaced providing the logic of the new test is consistent with the existing program. A possible example may be the future substitution of apoprotein A measurement instead of High density lipoprotein cholesterol (HDL-C).

The main interpretation takes place via the user mode which runs in either batch or interactive form, with or without a trace function. In the on line interactive mode patient information and first line test results (Total cholesterol, triglyceride, HDL-C and plasma storage test) are entered. A provisional diagnosis is made. This can be printed as an interpretative laboratory report. Further relevant medical information and drug history is requested and a full interpretative report is issued. The trace function projects a highlighted message of the rule name and action taken on the screen each time a rule is activated. The batch mode allows patient data to be run from a previously prepared file.

The system is currently undergoing exhaustive testing and evaluation using patient data from the lipid laboratory. It will undergo further testing in our lipid clinic in the near future. We expect that,as in our Thyroid Function Decision support system (5), the system is unlikely to produce incorrect results. The program is designed so that it will refrain from committing itself to a specific diagnosis unless the pattern of results is complete and clear. If any anomalies occur, the system may make a tentative suggestion but will always refer the user to the consulting physician/laboratory consultant for "human confirmation".

FUTURE ENHANCEMENTS

It is intended to incorporate a patient data base. This will allow adequate lipid clinic records to be maintained. The information in the data base can be used for statistical analysis or as a means to redefine both the reference ranges and decision levels in the light of user experience. (5) We also plan to expand the system to include diagnosis of patients with inherited (primary) hyperlipidaemias. This is important as some familial and secondary disorders respond to treatment differently. In the present system some patients with familial combined hyperlipidaemia and secondary type IV would both be classified as Fredrickson type IV.

A patient 'risk' predictor is also underdevelopment. This is based on the front line tests and age and sex of the patient. This may also include other personal patient details such as blood pressure, weight, smoking, drinking and exercise habits.

Finally we hope to allow the generation of new rules in the supervisory mode. This would enable the system to grow in parallel with advances in both laboratory and clinical medicine. The system has been designed to interface with other expert systems which we are developing. (5,6).

REFERENCES

1. The Lipid Research Coronary Primary Prevention Trial Results.
   JAMA 1984, 251, 351 - 374.

2. Consensus Conference on Lowering Blood Cholesterol to prevent Heart Disease.
   JAMA 1985, 253, 2080 - 2086.

3. Beaumont, J.L., Carlson, L.A., Cooper, G.R., Fejdart, Z., Fredrickson, D.S., and Strasser, T.
   Classification of Hyperlipidaemia and Hyperlipoproteinaemia.
   Bull. WHO, 1970, 43, 891 - 915.

4. Schaefer, E.J., and Levy, R.I.
   Pathogenesis and Management of Lipoprotein Disorders.
   N.E.J. Med. 1985, 312, 1300 - 1310.

5. Brosnan, P., Boran, G., Grimson, Jane B. and O'Moore, R.R.
   A Decision Support System for Assessment of Thyroid Function.
   Automedica 1987, (in press).

6. Al Zobaidie, A. and Grimson, Jane B.
   An Expert Clinical Data Base.
   MEDINFO (1986). Ed. Salamon, R., Blum, B. and Jorgensen, M.
   North Holland.

Lecture Notes in Medical Informatics